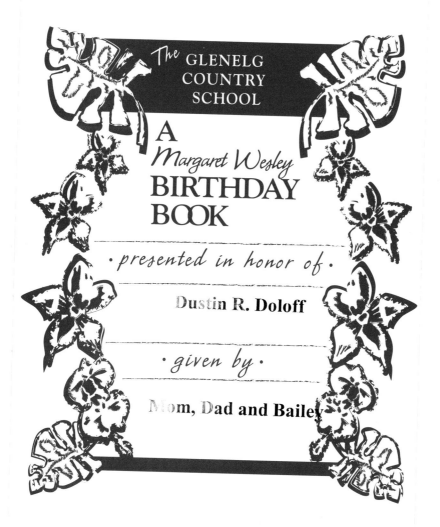

The GLENELG
COUNTRY
SCHOOL

A
Margaret Wesley
BIRTHDAY
BOOK

· *presented in honor of* ·

Dustin R. Doloff

· *given by* ·

Mom, Dad and Bailey

The Welfare of Horses

Animal Welfare

VOLUME 1

Series Editor

Clive Phillips, *Department of Clinical Veterinary Medicine, University of Cambridge, Cambridge, U.K.*

The Welfare of Horses

Edited by

Natalie Waran

University of Edinburgh,
Edinburgh, Scotland, United Kingdom

KLUWER ACADEMIC PUBLISHERS
DORDRECHT / BOSTON / LONDON

A C.I.P. Catalogue record for this book is available from the Library of Congress.

ISBN 1-4020-0766-3

Published by Kluwer Academic Publishers,
P.O. Box 17, 3300 AA Dordrecht, The Netherlands.

Sold and distributed in North, Central and South America
by Kluwer Academic Publishers,
101 Philip Drive, Norwell, MA 02061, U.S.A.

In all other countries, sold and distributed
by Kluwer Academic Publishers,
P.O. Box 322, 3300 AH Dordrecht, The Netherlands.

Printed on acid-free paper

Printed in the Netherlands.

TABLE OF CONTENTS

'ANIMAL WELFARE BY SPECIES' SERIES PREFACE

Animal welfare is attracting interest worldwide, but particularly from those in developed countries, who now have the luxury of being able to determine the management systems for their farm animals, as well has having plentiful resources for companion, zoo and laboratory animals. The increased attention given to animal welfare in the West derives largely from the fact that the relentless pursuit of financial reward and efficiency has lead to the development of intensive animal production systems that shame the conscience of many consumers in those countries. In developing countries, human survival is still a daily uncertainty, so that provision for animal welfare has to be balanced against human welfare. Welfare is usually provided for only if it supports the output of the animal, be it food, work, clothing, sport or companionship. In reality there are resources for all if they are properly husbanded in both developing and developed countries. The inequitable division of the world's riches creates physical and psychological poverty for humans and animals alike in all sectors of the world. Livestock are the world's biggest land user (FAO, 2002) and the population, particularly of monogastric animals, is increasing rapidly to meet the need of an expanding human population. Populations of animals managed by humans are increasing worldwide, so there is the tendency to allocate fewer resources to each one.

The intimate connection between animal, stockman and consumer that was so essential in the past is rare nowadays, having been superseded by technologically efficient production systems. Animals on farms and in labs are tended by fewer and fewer humans in the drive to increase labour efficiency. Consumers rarely have any contact with the animals that produce their food. In this estranged, efficient world man struggles to find the moral imperatives to determine the level of welfare that he should afford to animals within his charge. Some aim for what they believe to be the highest levels of welfare provision, such as the owners of pampered pets, others deliberately or through ignorance keep animals in impoverished conditions or even dangerously close to death. Religious beliefs and directives encouraging us to care for animals have been cast aside in a supreme act of human self-confidence, stemming largely from the accelerating pace of scientific development. Instead, today's moral code derives as much from horrific tales of animal abuse portrayed in the media and the assurances that we receive from supermarkets that animals used for their products were not abused in this way. The young were always exhorted to be kind to animals through exposure to fables whose moral message was the benevolent treatment of animals. Such messages are today enlivened by the powerful images of modern technology, but essentially alert children to the wrongs associated with animal abuse.

This series has been designed to provide academic texts discussing the provision for the welfare of the major animal species that are managed by humans. They are not detailed blue-prints for the management of animals in each species; rather they describe and consider the major welfare concerns of the species, often in relation to similar species or the wild progenitors of the managed animals. Welfare is considered in relation to the animal's needs, concentrating on nutrition, behaviour, reproduction and the physical and social environment. Economic effects of animal welfare provision are considered and key areas requiring further research. The requirements of different species are quite distinct and it is not possible to examine them in any detail in general texts covering groups of species linked by a common purpose, such as farm animals, except in a very superficial way. Hence by focussing on just one species in each book, and a few closely related genotypes, it is possible to consider the animals' needs in detail and how these can best be provided in today's society. Only by concentrating on one species has it been possible to consider the management of members of the species in a number of different situations, such as varied geographical locations, different economic climates etc. With the growing pace of knowledge in this new area of research, it is hoped that this series will provide a timely and much-needed set of texts for researchers, lecturers, practitioners, and students. My thanks are particularly due to the publishers for their support, and to the authors and editors for their hard work in producing the texts on time and in good order.

Clive Phillips,
Series Editor,
Department of Clinical Veterinary Medicine, University of Cambridge,
Cambridge, UK

Reference

Food and Agriculture Organisation (2002). http://www.fao.org/ag/aga/index_en.htm.

PREFACE

In 2000 there were an estimated 58.8 million horses in the world (FAO 2000). This figure represents a decline in numbers from 96.4 million in 1938 (immediately before the Second World War), when agricultural practises were far less mechanised than they are now. Although the largest proportion of the world's horses are still used for traction and draught purposes this is likely to continue to decline as agricultural practises in the developing world change. The highest proportion of the world's population of horses within the western or developed world, is found in the United States (5.3 million). In Europe, Germany has the greatest numbers (476,000).

The main role of the horse within the developed world is as a recreational or sports animal. In the United States, the horse industry was estimated to be worth 15 billion US dollars in 1990 (see Ensminger 1990), and horse racing ranked third behind baseball and car racing as a spectator sport. Horses are increasingly being kept as pets or companion animals, often by those with no previous knowledge about their physical or behavioural requirements. Yet despite this, until relatively recently, very little research was directed towards investigating the welfare of the domestic horse.

Welfare is about 'quality of life'. Scientists have attempted to define this concept so that it may be objectively assessed, but ultimately scientific evidence must be interpreted subjectively. There are no simple, universally accepted scientific indicators of welfare. Welfare cannot be measured, and assessments based upon measurements of behavioural, physiological and immunological responses to environmental challenges, are accepted as the best indirect evidence we have of an animal's subjective state. Recently more direct measurements of an animal's needs have been developed. Here the animal is offered choices between environments and conditions, and the strength of the preference can be measured. This provides scientists with a method that takes a more animal centred approach (anthropocentric) in determining the resources that are of importance in ensuring a good quality of life.

The problem is that 'quality of life' is often difficult to judge, even when the behavioural and physical requirements of animals have been determined. In addition, a particular individual's perception of a situation may differ from another's, due for example, to its previous life experiences. This is particularly pertinent when discussing the welfare of those horses used as work animals in under-developed countries. It is often perceived that they experience working conditions and management that fall well below the standards we, in the developed world, might expect for the sports or pet horse. By contrast, it is often assumed that the sports horse enjoys better housing and working conditions, yet the intensive management of the performance horse may predispose it to a different range of problems. In other

ix

words, the perceived welfare of horses kept for different purposes may differ from their actual welfare.

The aim of this edited book, is to provide the knowledgeable reader with a review of the current state of scientific research on the welfare of the horse. We hope that this will be used as an academic text, a reference book and as interesting reading for those who work with horses. To achieve this, the book begins with a discussion of the natural behaviour of the horse, and the effects of domestication. It then covers general issues that impact upon welfare, such as; nutrition, performance related clinical problems, housing and specific problems associated with intensive management, transportation and training methods. The book closes with two chapters in which the specific problems associated with the use of the horse either for sport or for work purposes are discussed.

In editing this book, I have learned a great deal and I have been surprised and encouraged by the extent of the recent work on horse welfare. I am extremely grateful to all of the authors of the chapters in this book. Their hard work and concern for horse welfare will be obvious to all those who read it. In addition I am grateful to all of my students. Their patience and their help with other aspects of my workload, reading chapters or providing a 'sounding board', has been invaluable. I am especially grateful to; Avanti Mallapur, Arnja Rose Dale, Shirley Seaman, and Anna Price. Sally West and Dot Marshall, are both thanked for their unfailing support at work and home respectively, especially during the editing process. I am also grateful to the series editor, Dr. Clive Phillips who was kind enough to encourage me to start this and also to carry out the final edit on all of the chapters.

Finally I want to thank my family, Chris, Kal and Conor, since without their tolerance, understanding and good humour, nothing would be possible.

Dr. Natalie Waran (2002)

References

Engsminger ME (1990) *Horses and horsemanship* (sixth edition). Interstate Publishers Inc., USA.
FAO (2000) http://www.fao.org.

LIST OF CONTRIBUTORS

Rachel A. Casey, BVMS, MRCVS
Anthrozoology Institute,
School of Biological Sciences,
University of Southampton,
Bassett Crescent East,
Southampton, Hampshire,
UK.

Deputy Director Anthrozoology Institute, University of Southampton. Currently funded by the Cats Protection, Rachel's research interests include the applied ethology of the cat, dog and horse, particularly with respect to clinical manifestations of behavioural change.

Andrew F. Clarke
Head of Equine Centre,
University of Melbourne,
Faculty of Veterinary Science,
Melbourne,
Australia.

Prof. Clarke is currently Head of the Equine centre, and chair of equine studies at The University of Melbourne's Faculty of Veterinary Science. The Equine centre has recently undergone extensive redevelopment, and over 3000 horses are referred there each year. Prof. Clarke is internationally known particularly for his research into the prevention and management of respiratory conditions in the horse. He is the author of numerous scientific papers and book chapters on this subject.

Jonathan Cooper, BA, PhD
Animal Behaviour, Cognition and Welfare Research Group,
University of Lincoln,
Lincolnshire School of Agriculture,
Caythorpe Campus,
Lincolnshire NG32 3EP,
UK.

Senior lecturer in Animal Behaviour and Welfare and Programme Leader for Degree in *Animal Management and Welfare*. Currently working on the behaviour and welfare of laying hens, the aquatic needs of ducks and stereotypic behaviour in stabled horses. Research interests include; development of abnormal behaviour, environmental enrichment and behavioural priorities.

Nell Davidson, BSc(Hons)
Equine Studies Group,
Waltham Centre for Pet Nutrition,
Freeby Lane,
Waltham-on-the-Wolds,
Leicestershire, LE14 4RT,
UK.

Equine Research Manager, WALTHAM centre for pet nutrition. At WCPN, Nell spent the first three years investigating aspects of the feeding behaviour of the domestic cat and dog before being recruited into the Equine research team (the Equine Studies Group). This role includes responsibility for managing the Equine Research Programme, particularly developing the Equine Behaviour Research Programme, with a specific goal of optimising the management of the domestic horse in today's society, to benefit the horse and the owner.

David L. Evans, BVSc, PhD
Faculty of Veterinary Science,
Gunn Building B19,
University of Sydney,
NSW 2006,
Australia.

Associate Professor, Faculty of Veterinary Science, University of Sydney, Research/ consultancy interests include horse training and fitness testing, equine pulmonary function and the epidemiology of lameness in racehorses.

Ted H. Friend, PhD, PAS, Dpl ACAABS
Department of Animal Science,
Texas A&M University,
2471 TAMU,
College Station, TX 77843-2471,
USA.

Professor, Department of Animal Science, TAMU. Teaches undergraduate and graduate courses and conducts research in applied animal ethology and has published research articles on all species of livestock and several exotic species including circus elephants and tigers.

Deborah Goodwin, BSc, PhD
Department of Social, Health and Behavioural Studies,
New College,
University of Southampton,
Southampton SO17 1BG,
UK.

Lecturer in applied animal behaviour, Department of Social, Health and Behavioural Studies and Deputy Director of the Anthrozoology Institute, University of Southampton. Research Interests include the behaviour of the domestic horse, cat and dog and their ancestral species. For example the effects of domestication on behaviour and the impact of humans on animals.

Pat Harris, MA, PhD, VetMB, MRCVS
Equine Studies Group,
WALTHAM Centre for Pet Nutrition,
Waltham-on-the-Wolds,
Melton Mowbray,
Leicestershire, LE14 4RT,
UK.

Equine nutritionist, WALTHAM Centre for Pet Nutrition and Adjunct Professor of Equine Studies at Virginia Polytechnic Institute and State University. Currently responsible for the equine research portfolio at WALTHAM and working closely with the Winergy (r) brand, Pat is a Former president of the British Equine Veterinary Association (1999), and serves on the executive board of Equine Veterinary Journal Ltd as well as on the Farriers Registration Council.

Des Leadon, MA, MVB, MSc, FRCVS
Irish Equine Centre,
Johnstown, Naas,
Co. Kildare,
Ireland.

Head of the Clinical Pathology Unit, Irish Equine Centre. RCVS registered Consultant/Specialist in Equine Medicine. Des is has an international reputation for his expertise in the transport of horses by air.

Paul McGreevy, BVSc, PhD
Faculty of Veterinary Science,
Gunn Building (B19),
Regimental Crescent,
University of Sydney.
NSW 2006,
Australia.

Senior lecturer in animal behaviour, Faculty of Veterinary Science, University of Sydney. Research interests are in horse behaviour and welfare especially the development of unwelcome behaviours such as stereotypies. Author of various papers on equines including a well known book called, Why does my horse . . . ? Souvenir Press.

Daniel S. Mills, BVSc, MRCVS
Animal Behaviour, Cognition and Welfare Group,
University of Lincoln,
Lincolnshire School of Agriculture,
Caythorpe Campus,
Lincolnshire NG32 3EP,
UK.

Principal Lecturer in BSc Animal Behaviour Studies & Animal Welfare, University of Lincoln and Director of Lincoln University Animal Behaviour Referral Clinic Daniel is a leading international authority on the subject of animal behaviour therapy. He combines both his behavioural and clinical knowledge to offer a referral clinic for animals with behaviour problems. Research Interests include; assessment of requirements of animals for good welfare, companion animal (including equine) problem behaviour and welfare, comparative psychiatry, individual differences and veterinary psychopharmacology.

Natalie Waran, BSc, PhD
Department of Veterinary Clinical Studies,
Royal (Dick) School of Veterinary Studies,
University of Edinburgh, Easter Bush,
Roslin, Midlothian EH25 9RG,
UK.

Senior lecturer in applied animal behaviour and animal welfare and Director of the Masters course in applied animal behaviour and animal welfare. Research interests include the behaviour of horses during transport by road and under intensive housing conditions. Recent work includes the assessment of pain in horses.

R. Trevor Wilson, DSc (Tropical Animal Production),
PhD (Ecology and Biology)
Bartridge House,
Umberleigh,
North Devon EX37 9AS,
UK.

Farmer in North Devon breeding pedigree Devon 'Red Ruby' cattle and Exmoor ponies. International consultant in agriculture. Worked with Shire horses on mixed farm after leaving school in 1954. Lifelong interest in equids and especially for draught and transport. Several peer-reviewed papers on Equidae.

Chapter 1

HORSE BEHAVIOUR: EVOLUTION, DOMESTICATION AND FERALISATION

D. GOODWIN
Research Fellow, Biodiversity and Ecology& Deputy Director,
The Anthrozoology Institute, University of Southampton, Bassett Crescent East, Southampton,
SO16 79X, UK

Abstract. The evolution of the horse began some 65 million years ago. The horse's survival has depended on adapative behaviour patterns that enabled it to exploit a diverse range of habitats, to successfully rear its young and to avoid predation. Domestication took place relatively recently in evolutionary time and the adaptability of equine behaviour has allowed it to exploit a variety of domestic environments. Though there are benefits associated with the domestic environment, including provision of food, shelter and protection from predators, there are also costs. These include restriction of movement, social interaction, reproductive success and maternal behaviour. Many aspects of domestication conflict with the adaptive behaviour of the horse and may affect its welfare through the frustration of highly motivated behaviour patterns. Horse behaviour appears little changed by domestication, as evidenced by the reproductive success of feral horse populations around the world.

1. Introduction

Most standard texts on horse behaviour present the horse as a social prey species, which survives by fleeing from predators. Although this is undoubtedly true for some horses, at some times, in some locations, this tends to disregard many important features of horse behaviour. Over simplification of the definition of 'normal' behaviour risks some aspects of equine behaviour being labelled as abnormal when in fact they are normal, adaptive and have ensured the survival of the species for 65 million years.

Most standard texts on horse behaviour present the horse as a social prey species, which survives by fleeing from predators. Although this is undoubtedly true for some horses, at some times, in some locations, this tends to disregard many important features of horse behaviour. Over simplification of the definition of 'normal' behaviour risks some aspects of equine behaviour being labelled as abnormal when in fact they are normal, adaptive and have ensured the survival of the species for 65 million years.

When the horse was domesticated some 6000 years ago, humans began taking horses from environments in which they had evolved, and managing them under conditions which were convenient to humans. Six thousand years ago at Dereivka in the Ukraine (Levine 1999) domestic horses were initially maintained as a food source in herded groups within their natural environment. However, as the role of the horse in human culture changed and diversified, the constraints of domestica-

1

N. Waran (ed.), The Welfare of Horses, 1–18.
© 2002 *Kluwer Academic Publishers. Printed in the Netherlands.*

tion began to restrict many aspects of horse behaviour. Today we may restrict horses' freedom to roam and their freedom to choose food, shelter, mates and social companions, depending on the purpose that horses are kept for. We also expect horses to accept and interact with us, and to understand our instruction, even though we evolved as predators and our own behaviour has been shaped by a very different evolutionary history.

Therefore, if we are to begin to understand why the domestic horse behaves in the way it does, we must first understand something of its evolutionary history and how horse behaviour is adaptive in diverse and changing environments. The direct ancestor of the domestic horse is generally believed to be extinct, probably due in part to human predation. Dwindling prey numbers may also have prompted the initial domestication process. However, evidence that the behaviour of the horse has changed very little during 6000 years of domestication is provided by the success of many feral populations of horses around the world. Feral horse populations can provide information about many aspects of adaptive equine behaviour, *e.g.* social behaviour, mate choice and reproductive behaviour, habitat selection and foraging behaviour.

The way that some of the constraints imposed on contemporary domestic horses conflict with this behaviour will be dealt with by authors of later chapters.

2. Evolution

The generally recognised ancestor of the earliest equids existed in the Eocene 65 million years ago. *Hyracotherium*, often known as *Eohippus* the Dawn Horse, was a primitive perrisodactile ungulate about the size of a fox, which ran on four toes on the front feet and three on the hind feet. It was a browser, with small low crowned teeth and inhabited swampy regions in what is now Wyoming in North America. Its multi-toed feet were well adapted for locomotion in this marshy environment.

Most diagrams of equine evolution are so simplified that they fail to represent just how successful the predecessors of our modern horses were. At their maximum diversity in the fossil record there were some 13 Equid genera, of which *Equus* is the only surviving genera today, which included some 30 separate species (MacFadden 1994). Today there are just seven extant *Equus* species remaining.

There is considerable diversification in the fossil record during the Eocene as early equids began to exploit a range of new habitats. By the Oligocene, *Mesohippus* and *Miohippus*, had achieved the size of large dogs, and ran on three toes. It is not until *Parahippus* appeared in the Miocene period that adaptations for life as a grazer on the plains begin to appear in the feet and teeth. The lateral digits still carried digital pads, but were unlikely to touch the ground unless travelling on very soft earth, or cornering at speed. *Pliohippus* was the first equid to lose the lateral digits completely, just leaving the metacarpals as long thin vestiges, which are further reduced in *Equus* to short splint bones, though three toed horses are occasionally born today (MacFadden 1994).

Behaviour, being a transient expression of activity, doesn't generally fossilise very well. However, due to the abundance of equids in the fossil record and especially at some sites *e.g.* the Owyhee Desert in Navada, where many individuals are preserved together, paleoethologists have begun to piece together information about the social structure and behaviour of these early equids. Fossils from such sites have provided evidence that early equid species showed adaptations in their population dynamics and behavioural ecology, which allowed them to exploit new and changing environmental resources. This trait may be viewed as a pre-adaption to domestication.

3. Social Organisation

Two general types of social organisation are recognised in extinct and extant equid species (Waring 1983; MacFadden 1994). Type I behaviour is seen today in the domestic horse, Przewalski horse, Burchell's zebra and the Mountain zebra and is characterised by a non-territorial family band of one stallion and up to six mares. Type I behaviour seems to be an adaptation to unpredictable environmental conditions and a regularly changing but constant food supply which may prompt migration. Type II behaviour is found in the domestic donkey, Grevy's zebra, African and Asian wild asses. Males are territorial and adults do not form lasting bonds. Females may range over several males' territories and will accept matings from any of these territory holders. This appears to be an adaptation to predictable but marginal semi-desert conditions.

An example of the adaptability of equine social organisation is provided by a population of feral horses on Shakleford Banks, off the coast of North Carolina in the USA where there is a population of territorial domestic horses (Rubenstein & Hohmann 1989). The island is sandy, about 11 miles long, less than 1 mile wide, and provides a very marginal habitat with high mortality rates within the population. On half of the island the horses adopt the usual non-territorial strategy, but in the eastern half of the island, where 2/3 of the island's 90 horses live, stallions actively defend territories. Access to very limited resources *e.g.* fresh water is so important that this population has adopted territoriality as a survival strategy. This is an exception to the norm, but it does demonstrate the flexibility of equine behaviour. This capacity for flexibility has played an important role in adaptation of the horse to the confines of the domestic environment.

4. Domestication

The earliest evidence for horses being associated with human culture comes from cave paintings made in France and Spain around 15,000 years ago when they were hunted for meat and hides. Around 9000 years ago the remains of wild horses become increasingly rare in archaeological sites in Europe. Around 6000 years ago the earliest evidence for the domestication of the horse begins to appear in the Ukraine, Egypt and western Asia. The first domestic equids may have been used

Table 1. Social organisation of extant Equid species

	Species	Territorial	Social group	Dispersal
TYPE I	Domestic Horse (*E. caballus*) Przewalski Horse (*E. przewalski*) Burchell's Zebra (*E. burchelli*) Mountain Zebra (*E. zebra*)	NO (Males defend harem)	Stable harem groups (Generally 1 male and multiple females)	Sub-adult males leave to join/form bachelor groups. Sub adult females join/form new harems.
TYPE II	Domestic Donkey (*E. asinus*) African Wild Ass (*E. africanus*) Asian Will Ass (*E. hemionus*)	YES (Males defend territories)	No lasting adult bonds (Females range over several male)	Sub-adult males join-form roaming bachelor groups. Sub-adult females range over several male territories.

as pack animals, then to pull sledges, and eventually wheeled vehicles (Clutton-Brock 1992). However, archaeological evidence from Dereivka suggests that horses were being ridden there at least 500 years before the wheel was invented (Levine 1999).

Until the end of the eighteenth century there were two subspecies of wild horse in Europe and Asia, the Tarpan (*Equus ferus ferus*) in Central Europe and the Przewalski (*Equus ferus przewalski*) or Mongolian wild horse. Both are now extinct in the wild, but an attempt to reconstruct the Tarpan from domestic hybrids has been made in Poland. The Przewalski is maintained as a captive population now numbering around 1300 individuals, some of which have been released into semi-natural reserves in Mongolia (Clutton-Brock 1992).

Whether the Przewalski or the Tarpan can be claimed to be ancestors of the domestic horse is debatable. It is possible that, like the domestication of the dog from the wolf, domestication of the horse happened in several places throughout the ancient world, and local subspecies of wild horse will have contributed to domestic stock. The contribution of the Tarpan to the Konik in Poland is an example of this, as is the contribution of the Przewalski to the Mongolian ponies.

An alternative hypothesis is that all of the horses in the world today are descendants of those domesticated at Dereivka 6000 years ago. The benefits derived from domestic horses may have been so great to human culture that this single domestication event spread rapidly through the human population, as occurred following the domestication of the cat in Egypt 4000 years ago.

By the first millennium BC the importance of the ridden horse in human culture had been established. Although asses and hybrids were ridden and used as beasts of burden in ancient and classical history, horses became preferred as war mounts in ancient Greece and then Rome, as they could be ridden behind the withers, which was much more comfortable and secure than the donkey seat (on the hind quarters), especially at speed. The role of the horse has mirrored the changes in human society ever since, as warhorse, draft horse and today as a sporting and companion animal. Though human society has changed rapidly, these changes have taken place in a very short period of evolutionary time. The fact that the behaviour of the horse has been little changed by domestication is evidenced by the ease at which it assumes a feral lifestyle, even in very marginal environments *e.g.* the Namib Desert, where there is a breeding population of horses descended from horses abandoned by soldiers at the end of the Second World War.

5. Breed Differences in Behaviour

It is possible that the diversity of type and behaviour within the domestic horse could pre-date domestication to some extent, if the multiple domestication event hypothesis is accepted. This could in part account for variations in the morphology and behaviour of northern and southern breeds.

Northern breeds (cold bloods/trotters/drafts) are generally heavier built, with deep bodies, short stocky legs, small ears, large heads, thick coats and less reactive temperaments (see Figure 1). These are all adaptations for energy conservation and survival in a cold climate.

The southern breeds, (the hot bloods/gallopers) are gracile with long slender legs, fine coats, small heads, large ears and other physiological adaptions to aid heat dissipation (see Figure 2). They are fast, highly reactive and enduring and are

Figure 1. Exmoor ponies show behavioural and morphological adaptations to energy conservation in the cold and exposed environment of Exmoor.

Figure 2. This Anglo-Arab endurance mare and her foal show physiological and morphological adaptations to heat dissipation in hot environmental conditions.

adapted to life in a hot arid environment. They are possibly best represented by the Jaf, or Persian Arab.

This cline in geographically remote members of the same species is described by Bergman's Rule and Allen's Rule (Moen 1973). Similar variation is seen in other mammals with a large geographic range *e.g.* the Timber wolf and the Ethiopian wolf, the Scottish wild cat and the African wildcat.

Unpublished data (Whitmore *pers. comm.*) on blood typing and skeletal remains suggests the existence of a third ancestral type. These are the gaited breeds of which the Moroccan Barb is a good example, though very few purebreds remain. These breeds are adapted to high altitudes, and gaiting is a safe fast way to move on scree, as one or more feet always remain on the ground. Other adaptations include a long back, very sloping croup, slightly sickle hocks, long neck and a large head with a straight or convex profile. All features which are far from aesthetically pleasing to the northern Europeans. The Moroccan Barb's conformation coupled with their belligerent temperament may explain why many attempts to 'improve' them by crossing with Arabian horses have been made. However, it appears that a swathe of indigenous gaited breeds and their descendants stretch across the Eurasian landmass from Spain and North Africa to Tibet. Examples include the Andalucian and Lusitano of the Iberian Peninsular, the Skyros pony of the Greek islands, the Indian Marwari and Kathiawari and the ancestral Akhal Teke.

6. The Human-Horse Relationship

During the period that the horse has been domesticated there have been two basic approaches to the human-horse relationship (Goodwin 1999). One approach is similar to that employed during the domestication of the dog, where humans attempt to establish their dominance over the horse. This approach seems to foster a desire to identify signals of submission in the horse by some trainers, and some equestrian disciplines judge the horse-human relationship by the amount of submission shown by the horse to the rider.

The other approach is co-operative based on an understanding of the behaviour of the horse. Both approaches were apparent during the ancient and Classical period. The Scythians and Greeks were observers of horse behaviour and employed their understanding of the behaviour of the horse in management and training. The Romans did not appear to inherit the skills of horsemanship from the Greeks, and many employed force and coercion in horse training. The Christian concept of Man's dominion over the beasts continued the decline in the treatment and training of the horse. Horses were trained using force and punishment and the idea of the cessation of pain, being a reward, became established and continues in some equestrian traditions today (Barclay 1980).

During the 18th Century the Duke of Newcastle, amongst others, began to revive the principles of Greek horsemanship and a number of famous equestrian schools were established which heralded the beginning of a return to more co-operative horse training. Both approaches persist to varying extents in different traditions of equitation today. However, the numbers of young, healthy horses 'breaking down' during training, or being slaughtered for behaviour problems, suggest that present day horse management and training has much room for improvement (Odberg & Bouissou 1999).

7. Behavioural Ecology of Feral and Free-Ranging Horses

Though the horse remains physically and mentally adapted to life on an open plain or mountain, it can adapt to other environments, such as the woodlands of the New Forest, or the marshlands of the Camargue. Feral and free-ranging horses generally occupy a home range and will attempt to return to it if moved through human intervention (Russel 1976; Berger 1986). Tyler (1972) estimated that the home range of New Forest mares was 82–1020 ha and reported that their location varied little between years. Each home range contained a grazing area, shelter, water and a 'shade' where ponies congregate to avoid the attack of biting flies (Duncan & Vinge 1979) and to reduce energy expenditure through heat stress in the summer months (Joubert 1972). Movements to shading areas increase summer range sizes in the New Forest compared to the winter ranges.

Horses are preferential grazers but also browse on a wide range of forbs, sedges, shrubs and trees. Feral and free-ranging horses eat for up to 16 hours per day and forage in grasslands and other habitats containing a range of vegetational communities (Hansen 1976; Putman *et al.*, 1987). The diet of New Forest ponies shows changes through the year with seasonal abundance and primary productivity of forage species (Gill 1988). In the summer, grasses constitute 60% of the diet of New Forest ponies, but this drops to 30% in the winter, when ponies forage in more sheltered habitats provided by gorsebreaks and deciduous woodland. During the winter, gorse, heather, forbs, shrubs and holly form most of their diet.

Eliminative behaviour differs between domestic and free-ranging horses (Carson & Wood-Gush 1983b). Domestic horses confined to pasture exhibit an aversion to grazing near faeces that results in the development of separate grazed and latrine areas (Ödberg & Francis-Smith 1977). Welsh (1973) reported that Sable Island horses eliminated indiscriminately in their home ranges with the exception of marking behaviour. This has also been reported for a bachelor group of Przewalski horses in a semi-natural reserve (Goodwin & Redman 1997), which suggests that latrine use may be an adaptation to domestication, possibly as a parasite avoidance mechanism (Burton 1992). Free-ranging New Forest ponies defecate in latrine areas within grassland areas on the Forest, though their use is less pronounced during the winter when grass productivity is at a minimum during the year (Putman *et al.*, 1991). It appears, therefore, that the use of latrine areas in domestic horses shows a dynamic relationship between animal density and the availability of clean grazing.

8. Social Behaviour

The social behaviour of the horse contributes to group stability and social affiliations are essential to systems of collaborative behaviour such as social facilitation, which influences communal activities (Fraser 1992). In the wild, membership of a group is such an important anti-predator and therefore survival device that the social behaviour of the horse functions to minimise conflict within the group and so promotes its stability.

Horses readily form social order within their groups and overt aggression in feral horse bands is relatively rare, compared with horses in the domestic environment (Houpt & Keiper 1982). Circular dominance systems are common in horses. Dominance order is unidirectional, but may not be linear throughout the group, so that A may be dominant to B who may be dominant to C, but C may be dominant over A (Houpt *et al.*, 1978).

The group order may therefore be complex, but stable within stable populations, and though dominant animals may have been aggressive in the past, they do not need to be aggressive subsequently in all social situations in order to maintain their position. Once established the relationships of all horses within the group persist for as long as the group remains a closed unit.

In feral and free-ranging horses unsettled dominance relationships are usually only found between young horses, and free-ranging equine society could be said to function on kinship, recognition and respecting another's space. Dominance in horses is, therefore, related to control of space and avoidance of conflicts. In view of this the avoidance order is a better measure of the social system, than the aggression order (Fraser 1992). The avoidance order can become unstable if space is restricted, as happens in many domestic environments. The operation of an avoidance system is the key to understanding the social behaviour of horses. It is both obvious and subtly obscure, and so has been overlooked by many concerned with aggression and dominance in understanding the behaviour of the horse.

Social dominance is sometimes actively exerted by dominant individuals in feral groups, but is seen far more frequently in the domestic environment in competition for limited resources such as supplementary feed, or access to water troughs. This explains the high frequency of aggressive interactions recorded in domestic horses (Crowell-Davis 1993). But even here given adequate space, subordinate individuals will avoid moving too close to dominant ones. Another problem associated with the domestic environment, and particularly in livery yards (UK) and Barns (USA), is that the membership of the social group is constantly changing, and therefore never stable. This results in high levels of aggression and resultant injury, though there are ways of trying to minimise this, such as introducing newcomers gradually and preferably in pairs which have been previously accustomed to each other.

9. Maintenance Behaviour

Maintenance behaviour e.g. stretching and grooming, is exhibited by horses of all ages and play an important role in contributing to the well being of the horse (Waring 1983). Its importance may be easily overlooked as it may be short and varied, but these activities are repeated frequently through the day. Self-grooming includes rolling, rubbing and scratching, as well as licking and nibbling the coat. The face is groomed by rubbing against objects or the inside of a foreleg.

The horse needs around two hours of recumbent sleep during each 24 hour period though this may be taken in several short bouts due to the pressure exerted on the viscera. Horses have a unique stay apparatus in their hind legs which enables them to sleep or drowse on their feet for an additional five hours per day, and this is usually taken in relatively short bouts. Individuals take turns to sleep within a group while others remain alert as a predator avoidance strategy (Fraser 1992).

10. Predator Avoidance

Like many social prey species, the horse's sensory systems and social behaviour have adapted to facilitate early detection of approaching predators. The horse demonstrates two main predator defence strategies, which depend on environmental conditions and the nature of the predator. There is also some evidence of breed differences in the type of strategy employed.

Predation by Pantherine predators, which commonly kill by leaping onto the back and applying a kill-bite the throat, appear to been associated with the evolution of the rapid flight response which keeps the vulnerable head area as far away from the predator as possible. This is characteristic of highly reactive breeds, *e.g.* the Arab, originating from hot arid areas where ground conditions are conducive to galloping and predation by big cats is probable.

An alternative predator defence strategy is directed towards canid predators and is common when ground conditions are boggy, or on mountain scree, when rapid flight may not be possible or would not be a good survival strategy due to the risk of injury. In these conditions horses will stand at bay and defend themselves with foreleg strikes, which are capable of smashing the skull of a dog or wolf. Iberian horses, the Moroccan Barb and many British native ponies employ this strategy.

Defence strategies against human predators have not been reported in the literature, but predator defence strategies must be acknowledged and avoided in human-horse interactions. Like most prey species, horses learn to distinguish hunting behaviour from other behaviour in constantly present predators. Fortunately another feature of anti-predator behaviour appears to have pre-adapted the horse to domestication, in that they generalise their behaviour to include other species.

Equids are social animals, preferring to associate with others of their own kind, though accepting other species as companions too *e.g.* zebra and wildebeest commonly associate on the African Savannah. As prey species, group living is an important survival strategy, as it increases the probability of detecting approaching predators, it also reduces the probability of any particular individual being caught and consumed. Domestic equids are, therefore, pre-adapted to forming associations with other species and to respond to the warning signals in the body language of other species (Goodwin 1999).

11. Matriarchal Society

Studies of feral and free ranging horses in the UK (Tyler 1972) and USA have shown that horse society is basically matriarchal and consists of stable associations between mares and their offspring (Wells & Goldsmidt-Rothschild 1979). These associations persist even in the absence of a stallion, as seen in some of the managed free-ranging populations of native ponies in the UK. These generally have very few stallions present, figures of one stallion to 60 mares have been recorded for ponies of the Gower peninsular in Wales. In the New Forest during the summer, ratios of one stallion to 30 mares are common (Gill 1988). During the breeding season, stallions may collect several family bands of mares together forming fairly large but temporary harems. In the past many of the stallions have been removed from the New Forest for the winter and the large harems disperse into their component small matriarchal groups again. Those stallions remaining on the Forest over winter generally associate with a preferred mare and her family group and do not attempt to maintain the large harems that they hold during the summer.

In feral unmanaged populations, harems or bands usually have a single stallion, though multiple stallion bands do exist. In these the dominant male secures most of the matings, but subordinate stallions do have access to some matings, making subordinate status more reproductively advantageous than life in a bachelor group

(Miller 1981). Bands occupy a familiar home range, the size of which varies with the availability of resources within it. Occasionally bands join together to form large herds, e.g. around a scarce water resource, as in the horses of the Red Desert in Wyoming. These large associations are generally temporary, but within them inter-band recognition and dominance hierarchies have been shown to exist, suggesting the presence of social structure even in these temporary large associations (Miller & Denniston 1979).

12. Stallion Behaviour

Interactions between harem stallions generally involve much posturing, squealing and displays of marking behaviour i.e. ritualised dunging. Stallions tend to dung repeatedly on heaps known as 'stud piles' (Tyler 1972) and studies have shown that the dominant individual is last to dung on a pile during marking display bouts (Carson & Wood-Gush 1983b).

As in most social species overt aggression is costly and most encounters are resolved before the escalation of display into actual conflict. Fights resulting in death of one of the opponents have been recorded during battles over the possession of harems. However, most conflicts end in stallions parting and herding their mares away with characteristic snaking movements of the head and neck (Waring 1983).

Despite the popular macho image of the stallion, equine family bands are generally led by mares, and a study by Houpt & Keiper (1982) showed that in feral and domestic horse groups, stallions were neither the dominant nor most aggressive animals in their herds, and that all stallions studied were subordinate to some of the mares in their groups.

13. Reproduction

Reproduction in horses is seasonal and the annual cycle of oestrus periods is related to day length. This can be artificially manipulated for some breeds *e.g.* the Thoroughbred (TB), but the majority of non-TB foals are born in the late spring and early summer months (Fraser 1992). This corresponds in temperate ecosystems with the peak period of vegetational productivity, providing optimum nutrition for the lactating mare, and ensuring maximum growing time for the foal in relatively mild climatic conditions before the onset of its first winter (Gill 1988).

Courtship under natural conditions begins several days before oestrus. The stallion actively discourages contact between the mare and other males. As the mare enters full oestrus the stallion will remain very close to her. He will begin to test her readiness to mate by licking and nuzzling her, which stimulates the mare to urinate and adopt a standing position with back legs straddled and tail raised and arched to one side. This posture is exhibited prior to full oestrus when the mare urinates if approached but at this stage the stallion's advances will be rejected with much squealing, tail lashing and stamping. The stallion will often exhibit flehmen in response to the mare's urine, and he will assess the mare's willingness to mate by nudging and nibbling at her flanks, hindquarters and hind legs. The stallion may also place his head on her rump before assuming the relatively vulnerable position behind her where he may be kicked if she is not ready to be mated. When full oestrus is reached and successful mating has taken place the pair will remain together for several days and mate repeatedly. This natural form of sexual bonding ensures the optimum conditions for fertilisation to occur (Waring 1983).

Gestation takes around 11 months and 80% of foals are born at night. Many mares leave their social group to give birth and can delay parturition if they are disturbed, presumably as an anti-predator device. The udder may become swollen and wax may accumulate on the teats several days before birth, but onset of the first stage of labour is usually indicated by sweating and restlessness around one to four hours prior to the birth. The final stages of labour are very rapid, and the majority of foals are delivered when the mare is recumbent. Expulsion of the foal can take as little as 10–30 minutes after which the mare continues to lie still for an average of 15–20 minutes. After this recovery period she will usually rise and begin to lick the foal for up to 30 minutes, during which time the mother-foal bond is established. Almost immediately after birth, foals begin to make struggling movements with the legs and to tilt the head upwards. After gaining its feet it begins making random teat-seeking movements, exploring the underside of the dam, and eventually finding the udder. To suckle the foal is required to adopt a parallel and opposite position with the mare and to extend and rotate the head and neck. To facilitate nursing the mare flexes the opposite hind leg tilting the pelvis, and therefore the udder, towards the foal (Fraser 1992).

14. Behavioural Development

Foals are precocial developers and can stand and suckle within two hours of birth. Those born at night in the open are generally capable of keeping up with their mothers by dawn, which is essential as foals follow their mothers, rather than lying in undergrowth as is the case for fawns or calves. In the first weeks of life foals stay very close to their mother's side but between one and two months of age they begin to make exploratory trips away from the mother, practising locomotory skills, and begin playing with other foals (Carson & Wood-Gush 1983a; Crowell-Davis 1986). From two to three months foals begin to spend much of their time associating with other foals and begin to form peer groups. Weaning takes place naturally at around 8–9 months, though some mares will continue to suckle their foal until shortly before the arrival of their next foal (Gill 1988). Foals in feral and free-ranging horses remain with their natal groups until they approach maturity. During this time young horses learn social and survival behaviour *e.g.* habitat and forage selection, which contribute to reproductive success in later life. In some managed free-ranging populations in the UK, *e.g.* the New Forest where stallion numbers are relatively low, some mares and their female offspring remain together throughout much of their lives.

As the age of the peer group increases, group activities change from play and rest to increasing amounts of grazing time. Foals move between peer groups and their mothers in a variety of activities (Crowell-Davis *et al.*, 1987). As a result the primary social bonding is in kinship groups at weaning. Within peer groups close pair bonds develop between individuals which can persist throughout life, particularly between mares. Pair bonded individuals associate closely together and appears to derive social support from the bond, particularly during agonistic interactions with other individuals. Bonded pairs exhibit many affiliative interactions including mutual grooming, neck overlapping and resting head-to-tail fly swishing. Bouts of mutual grooming also occur within peer groups and may initiate play bouts (Crowell-Davis *et al.*, 1986).

Play has a vital role in the development of the horse with up to 75% of the kinetic activity of foals devoted to it. Solitary and object play appear to develop within the first month of life when foals remain close to their mothers. Social play develops from around 4 weeks of age as foals begin to interact with their peers and other members of the social group (unpublished data, Hughes & Goodwin). Social play in peer groups may involve chasing games, and biting and wrestling which may escalate into low intensity agonistic fighting. Fraser (1992) argues that during the process of domestication, and during socialisation to humans the naturally high levels of movement-related and social play have been channelled into forms of work and recreational activities. He believes this explains how the horse has dealt competently with domestication, while preserving certain behavioural characteristics, such as reactiveness, which make it able to survive in the wild.

Young horses leave their natal groups as they approach maturity. The majority of fillies leave when they are between 1.5 and 2.5 years old when 80% join other existing harems, the remainder begin new ones with bachelor stallions. A study by Monard *et al.* (1996) has shown that fillies leave not because of intrasexual competition, since they are not treated aggressively by the mature mares of the band, but because they do not accept matings from males in the natal group. If sexual advances were made by the males in the natal group the dams of the young females were observed interrupting the male's advances. This lack of sexual interest in males of the natal group is seen in many social mammals, including humans, and is thought to be a means of avoiding inbreeding.

Juvenile colts leave, or may be expelled from their natal bands by the stallion, and form bachelor groups where they practise the skills necessary for the acquisition of their own band of mares. Solitary horses are rare and are usually deposed harem stallions, though these may join bachelor groups because of the companionship and safety afforded by them.

15. Communication

Horses are primarily visual communicators though vocal communication is also evident (Fiest & McCullough 1976). The meaning of individual sounds may be complex and context dependent. Vocal communication can serve to maintain contact over long distances, indicate excitement, deter contact in social interactions, or initiate approach between a mare and her foal. Odours are used for communication over time through marking behaviour, in courtship and in establishing the mare foal bond. Touch is used for communication at close range, it promotes and maintains pair bonds during mutual grooming and can be a form of social support in stressful situations. The latter may be derived from the foal pressing its body against the dam during novel or disturbing events.

Horses are extremely sensitive to subtle changes in the body language of their companions. As they generalise their communication to include us, they also react to our body language whether this is intentional on our part or not.

Body outlines are very important in equine communication. High rounded outlines indicate excitement and low straight outlines are associated with relaxation (Rees 1993). Interestingly these outlines also have a psychological effect on humans and the way horses and humans are portrayed in art. Kings or Emperors were often pictured mounted on horses displaying high rounded outlines, whilst people of a more lowly disposition are usually mounted on horses displaying a low outline.

The alarm posture of the horse serves to alert the herd to possible danger, and it is, therefore, a posture of high tension. This sensitivity in the horse to bodily tension is an aspect of their communication, which they generalise towards their communication with humans. Horses react to tension in humans with the same alarm that they would if exhibited by equine companions. They are so subtle in their perceptions that it can very difficult to hide feelings of human nervousness from them.

Many parts of the body can be used independently as signalling structures. The tail can be used to signal excitement, arousal, fear and aggression. Legs can deliver redirected threats. Ears can be moved independently and towards the direction of sounds of interest, they can indicate the direction of the horse's attention and its state of arousal.

Even more complex signals are portrayed by the face and can indicate relaxation, irritation, tension and even pleasure (Rees 1993). Relaxation is seen in drooping eyelids, ears and lips. Wrinkling around the mouth and nostrils indicates intense irritation, and may be displayed by some horses when confronted with a saddle, or when a disliked horse or person passes near to them. Tension is associated with a deep triangle that appears above the mouth, and the lips are held tight, often with a dimpled chin. A long nose without tension and half-closed eyes are often associated with the pleasure of grooming, either rubbing against a convenient object or when provided by a horse or human companion.

16. Agonistic Signalling

Like many social animals horses show escalated warnings of aggression, and due to the consequences of ignoring these they have received more attention in the literature than other aspects of equine communication and behaviour. Much of the aggression seen in domestic horses is ultimately due to management conditions and may have a variety of causes *e.g.* competition over small amounts of highly nutritious food, failure by handlers to respond to more subtle threat signals, or unrecognised pain.

The aggressive threat signals of the horse show a general pattern of escalation, which may begin with relatively mild nose wrinkling and ear flattening, and then proceed through head jerk, teeth bearing, tail lashing, stamping a foreleg and charging. Aggressive threats generally culminate in biting (Rees 1993). The general pattern can show individual variation and be complete or become truncated if appropriate avoidance is not shown. A direct attack without warning is usually a learnt response directed towards humans who have failed to respond repeatedly to more subtle equine threat signals.

Defensive threats begin with the same signals of nose wrinkling and ear flattening, possibly allowing the horse time to decide whether to react aggressively or defensively to an aggressor. They then escalate through blocking body movements, tail flattening and rump presentation, raising a hind leg, backing up and hind leg kicking (Waring 1983). It is interesting to note that defensive threats culminate in kicking, rather than biting, due to the reduced risk to the vulnerable head and neck.

In comparison there are few obvious signals of submission in the horse, submission appears to be expressed by moving away from a threat or desired resource, possibly with ears half back. It is possible that very subtle signals exist which biologists have failed to notice, but due to the horse's regard for ownership of space, it may be that acknowledged control of space within the herd makes signals which are obvious enough for humans to recognise unnecessary (Fraser 1992). There are a number of signals that some have claimed to indicate submission but on closer examination this may not be the case. Foals exhibit snapping or mouthing towards other horses which is characterised by drawing back the lips and snapping the jaws, with the head and neck outstretched. This is a common juvenile gesture and generally ceases to be expressed after puberty, though snapping persists in some adults. Snapping was originally considered to be submissive, but Crowell-Davis *et al.* (1985) observed that snapping failed to inhibit aggression from con-specifics and may even trigger aggression in some cases, therefore, if it is a submissive gesture, it is an inefficient one. Crowell-Davis *et al.* (1985) suggested that it may have multiple meanings depending on context, or that it is a displacement activity derived from nursing or grooming behaviour. However, as snapping also occurs during greeting behaviour of zebras (Schilder *et al.*, 1984) it may have a similar meaning in horses.

Licking, chewing, and head-lowering are considered important submissive signals by some horse trainers, who believe that licking and chewing are the adult form of snapping, which they consider to be submission. However, licking and chewing may be displacement activities exhibited when horses experience conflicting motivations. These signals have been reported by Houpt *et al.* (1978) to occur when horses are expecting food and they could be explained by conflicting desires of wanting to begin feeding, but having to wait for the approach of the human feeder. Alternatively, these oral movements may be self comforting following a period of anxiety and subsequent adrenalin production, which tends to dry the mouth, by distributing saliva over the mucosa of the mouth (McGreevy *pers. comm.*).

Isaac and Goodwin (1998) studied head-lowering and snapping in free-ranging New Forest ponies in the New Forest, Hampshire, UK. They reported snapping by New Forest foals during interactions with members of their natal group. Snapping was observed during aggressive and non-aggressive interactions, which agreed with earlier work, by Crowell-Davis *et al.* (1985). They also observed that head-lowering in New Forest ponies only took place during the approach phase of interactions, but was never observed in response to aggression. They concluded that head-lowering was, therefore, not a submissive signal but may instead be a distance-reducing and affiliative signal. Another example of a distance-reducing, affiliative signal is the 'Tail Up' signal in the domestic cat (Cameron-Beaumont 1997).

Isaac (1998) also reported that snapping was shown by adults and foals of the social Chapman's zebra during greeting behaviour and postulated that snapping in domestic foals may, therefore, also be a greeting behaviour which could explain why it induces both affiliative and aggressive reactions from conspecifics. Adult and foal Chapman's zebra also showed head-lowering during the approach phase of interactions. Adult and foal Grevy's zebras also exhibited head-lowering during the approach phase of interactions, but only Grevy's zebra foals showed snapping during interactions directed towards the herd stallion and this was associated with non-aggressive avoidance.

Although further study is necessary to gain a better picture of the form and function of submissive signalling by horses, this must be undertaken with due diligence as it may prompted more by a human cultural desire to recognise submission than by its importance in equine behaviour.

17. Conclusion

Whilst the horse has undoubtedly benefited from some aspects of domestication *e.g.* in the provision of food, shelter, protection from predators and care during illness and injury, many of the constraints imposed on domestic horses conflict with their evolutionary adaptive behaviour. The success of feral horse populations around the world indicates that the adaptive behaviour of the horse has changed very little in 6000 years of domestication. However, it can not be assumed that the feral condition equates with optimal welfare, as many feral and free-ranging populations survive in sub-optimal environmental conditions (Waran 1997).

The ability of domestic horses to adapt to some current intensive management regimes seems to be unacceptable for a rising percentage of the equine population. Stabling and restriction of foraging time have been associated with the development of abnormal behaviour in competition horses (McGreevy *et al.*, 1995). In a recent survey, up to 29% of leisure horse owners reported experiencing problem behaviour in their horses in relation to some aspect of their management and stabling (Sommerville *et al.*, 2001). Studies of feral and free-ranging horses can be used as indicators of behaviour that is important to horses and can be useful in determining possible causes of maladaptive behaviour in domestic and captive horses. It is also likely that domestic horse welfare could be improved by meeting the need to perform highly motivated behaviour identified from studies of feral and free-ranging horses.

18. References

Barclay, H.B. (1980) *The Role of the Horse in Man's Culture.* J.A. Allen, London, UK.

Berger, J. (1986) *Horses of the Great Basin.* Chicago University Press, USA.

Burton, D. (1992) *The Effects of Parasitic Nematode Infection on Body Condition of New Forest Ponies.* PhD Thesis, Department of Biology, University of Southampton.

Cameron-Beaumont, C.L. (1997) *Visual and Tactile Communication in the Domestic Cat (Felis silvestris catus) and Undomesticated Small Felids.* PhD Thesis, Division of Biodiversity & Ecology, University of Southampton.

Carson, K. and Wood-Gush, D.G.M. (1983a) Equine Behaviour II: A review of the literature on feeding, eliminative and resting behaviour. *Applied Animal Ethology* **10**, 179–190.

Carson, K. and Wood-Gush, D.G.M. (1983b) Equine Behaviour I: A review of the literature on social and dam-foal behaviour. *Applied Animal Ethology* **10**, 165–178.

Clutton-Brock, J. (1992) *Horse Power: A History of the Horse and Donkey in Human Societies.* Harvard University Press, Massachusetts, USA.

Crowell-Davis, S.L. (1986) Developmental behaviour. *Equine Practice* **2, 3**, 573–559.

Crowell-Davis, S.L. (1993) Social behaviour of the horse and its consequences for domestic management. *Equine Veterinary Education* **5, 3**, 148–150.

Crowell-Davis, S.L., Houpt, K.A. and Burnham, J.S. (1985) Snapping by foals. *Zeitschrift Fur Tierpsychologie* **69**, 42–54.

Crowell-Davis, S.L., Houpt, K.A., and Carini, C.M. (1986) Mutual grooming and nearest neighbour relationships among foals of *Equus caballus*. *J. Applied Animal Behaviour Science* **15, 2**, 113–123.

Crowell-Davis, S.L., Houpt, K.A. and Kane, L. (1987) Play development in Welsh pony foals. *J. Applied Animal Behaviour Science* **18**, 119–131.

Duncan, P. and Vinge, N. (1979) The effect of group size in horses on the rate of attacks by blood-sucking flies. *Animal Behaviour* **27**, 623–625.

Fiest, J.D. and McCullough, J.D. (1976) Behaviour patterns and communication in feral horses. *Zeitschrift Fur Tierpsychologie* **41**, 337–371.

Fraser, A.F. (1992) *The Behaviour of the Horse.* C.A.B. International, Wallingford, UK.

Gill, E.L. (1988) *Factors Affecting Body Condition of New Forest Ponies.* PhD Thesis, Department of Biology, University of Southampton.

Goodwin, D. (1999) The importance of ethology in understanding the behaviour of the horse. *Equine Veterinary J. Suppl.* **28**, 15–19.

Goodwin, D. and Redman, P. (1997) Eliminatory behaviour of a bachelor group of Przewalski horses in a semi-reserve: comparison with the domestic horse. In Hemsworth, P.H., Spinka M. and Kos L. (eds.), *Proceeding of 31st International Congress International Society for Applied Ethology*, p. 69. Research Institute of Animal Production, Prague, Czech Republic.

Hansen, R.M. (1976) Foods of free-roaming horses in Southern New Mexico. *J. Range Management* **29, 4**, 437.

Houpt, K.A., Law, K. and Martinisis, V. (1978) Dominance hierarchies in domestic horses. *Applied Animal Ethology* **4**, 273–283.

Houpt, K.A. and Keiper, R.R. (1982) The position of the stallion in the equine dominace hierachy of feral and domestic ponies. *J. Animal Ethology* **9**, 111–120.

Isaac, N. (1998) *A Study of Submissive Signals and Behaviour Patterns of Domestic and Wild Equid Species.* BSc. Honours project, Division of Biodiversity and Ecology, University of Southampton.

Isaac, N. and Goodwin, D. (1998) *Visual Communication in Horses and the Horse-Human Relationship.* International Society of Anthrozoology Conference, 1998, Prague, Czech Republic.

Joubert, E. (1972) Activity patterns shown by mountain zebra (*Equus hatmannae*) in South West Africa. *Zoologica Africana* **7, 1**, 309–331.

Levine, M.A. (1999) Investigating the origins of horse domestication. *Equine Veterinary J. Suppl.* **28**, 6–14.

MacFadden, B. (1994) *Fossil Horses*. Cambridge University Press, Cambridge, UK.

McGreevy, P.D., French, N.P. and Nicol, C.J. (1995) The prevalence of abnormal behaviours in dressage, eventing and endurance horses in relation to stabling. *Veterinary Record* **137**, 36–37.

Miller, R. (1981) Male aggression, dominance and breeding behaviour in Red Desert feral horses. *Zeitschrift Fur Tierpsychologie* **57**, 340–351.

Miller, R. and Denniston R.H. (1979) Interband dominance in feral horses. *Zeitschrift Fur Tierpsychologie* **51**, 41–47.

Moen, A.A. (1973) *Wildlife Ecology*. W.H. Freeman & Co., San Fransisco, USA, pp. 294–296.

Monard, A.M, Duncan, P. and Boy, V. (1996) The proximate mechanisms of natal dispersal in female horses. *Behaviour* **52, 3**, 1095–1124.

Ödberg, F.O. and Bouissou, M.F. (1999) The development of equestrianism from the baroque period to the present day and its consequences for the welfare of the horse. *Equine Veterinary J. Suppl.* **28**, 26–30.

Ödberg, F.O. and Francis-Smith, K. (1977) Studies of ungrazed eliminative areas in fields used by horses. *Applied Animal Ethology* **3**, 147–188.

Putman, R.J., Prat, R.M., Ekins, J.R. and Edwards, P.J. (1987) Food and feeding behaviour of cattle and ponies in the New Forest, Hampshire. *J. Applied Ecology* **24**, 369–380.

Putman, R.J., Fowler, A.D. and Tout, S. (1991) Patterns of use of ancient grassland by cattle and horses and effects on vegetational composition and structure. *Biological Conservation* **56, 3**, 329–347.

Rees, L. (1993) *The Horse's Mind*. Stanley Paul, London, UK.

Rubenstein, D.I. and Hohmann, M.E. (1989) Parasites and social behaviour of island feral horses. *Oikos* **55**, 312–320.

Russel, V. (1976) *New Forest Ponies*. David and Charles, London, UK.

Schilder, M.B.H., Hooff, J., Geer-Plesman, C.J. and Wensing, J.B. (1984) A quantitative analysis if the facial expressions in the Plains zebra. *Zeitschrift Fur Tierpsychologie* **66**, 11–32.

Sommerville, K., Goodwin, D., Bradshaw, J.W.S. and Lowe, S. (2001) *The Prevalence of Behaviour Problems in the Domestic Horse*. Companion Animal Behaviour Therapy Study Group Meeting 2001. 4th April 2001 Birmingham, UK.

Tyler, S.J. (1972) The behaviour and social organisation of New Forest ponies. *Animal Behaviour Monographs* **5, 2**, 85–196.

Waran, N.K. (1997) Can studies of feral horse behaviour be used for assessing domestic horse welfare? *Equine Veterinary J.* **29, 4**, 249–251.

Waring, G. (1983) *Horse Behaviour: the Behavioural Traits and Adaptations of Domestic and Wild Horses*. Noyes Publications, New York, USA.

Wells, S.M. and Goldschmidt-Rothschild, B. (1979) Social behaviour and relationships in a herd of Camargue Horses. *Zeitschrift Fur Tierpsychologie* **49**, 363–380.

Welsh, D.A. (1973) The life of Sable Island's wild horses. *Nature Canada* **2**, 7–24.

Chapter 2

CLINICAL PROBLEMS ASSOCIATED WITH THE INTENSIVE MANAGEMENT OF PERFORMANCE HORSES

R.A. CASEY*
Anthrozoology Institute, School of Biological Sciences, University of Southampton, Bassett Crescent East, Southampton, Hampshire, UK

Abstract. The physical as well as the behavioural requirements of the horse changed little through the process of domestication. This means that horses kept within an intensively housed environment and used for performance, physically and behaviourally are susceptible to specific clinical conditions, injuries and diseases. In this chapter, physiological and clinical problems such as those causing pain related behaviours and head shaking are discussed. The most commonly associated problems with horses kept in intensive housing conditions or used in specific competitive disciplines are highlighted. Despite the increasing amount of information about injury and disease in the horse, there is little research relating such problems to the situations performance horses have to cope with. This is particularly the case with pain, whose recognition of pain amongst professionals is still variable and often subjective and not widely recognised as a cause of behavioural change.

1. Introduction

Keeping horses within a stabled environment, and using them for riding and driving purposes does require a consideration about how these environments and activities affect the normal behavioural requirements and motivational drives in horses, as discussed in the previous chapter and in Chapter 5. However, it is also pertinent to consider that the conditions in which horses are kept and the activities for which they are used may affect their physical as well as psychological welfare. Pathological or painful conditions can arise where horses are kept in situations that are incompatible with their natural physiology, or can arise through injury associated with particular forms of activity or particular types of equipment. In this chapter, the effect of modern management and working or competition practices on the physical welfare of horses will be examined to suggest where particular practices may compromise the welfare of horses. The chapter is divided into problems that arise through management practices, such as those from stabling or grazing practices, and those which arise through riding, such as pain related disorders and headshaking. Obviously there is an enormous number of veterinary conditions that can impact on the welfare of horses – it is not the aim of this chapter to cover all of these, but to indicate areas of major concern, and highlight condi-

---

* Note: Daniel Mills, University of Lincoln, is thanked for his contribution on head-shaking.

N. Waran (ed.), The Walfare of Horses, 19–44.

tions that have a major impact in terms of the welfare of the modern competition equine.

2. Management Associated Welfare Problems

2.1. STABLING

The manner in which domestic horses are managed in the UK depends a great deal on the purpose for which they are kept, where and how they are kept, and the time of year (Harris 1999). Whether horses are kept in a traditional stable will depend on all of these factors, and it is tempting to generalise that, for example, racing horses and dressage horses spend the majority of their time stabled, and general riding horses spend a greater proportion of their time outside grazing. Although this may be true for a major proportion of the population of each type of animal, there is inevitably a huge variation in management factors between individuals within each activity type. As discussed in the previous chapter, keeping horses in a traditional stable often fails to provide them with an environment that fulfils their motivational drives. However, as well as inhibiting behavioural needs, stabling can also result in horses developing more physical forms of compromised welfare (see Figure 1).

Figure 1. Inappropriate foot-care can cause many clinical problems (Courtesy of The Donkey Sanctuary).

The natural diet for horses is a varied low quality roughage diet, which they are designed to eat for a significant proportion of each day in order to obtain sufficient energy for survival. The digestive system of the horse is rather less efficient than other herbivore groups such as the ruminants, and lagomorphs. The tough cellulose of grasses has led these species to develop digestive systems that allow for the breakdown of the cellulose cell walls to enable full digestion of the sugars inside. Ruminants chew digesta for a second time after partial digestion in the rumen, and lagomorphs use a system of a complete second digestion to digest the sugars released after caecal digestion of cellulose. The digestive system of the horse relies on a single passage of materials, and hence less energy is released from an equal volume of grass consumed. Horses, therefore, not only need more grass, of a better quality, to survive, but they also need to spend more time eating and digesting it. The modern practices used for stabled competition horses, whereby a high energy diet is provided in small hard feed rations does not match that for which the digestive system of the horse was designed. This can lead to abnormalities of digestive function, such as ulceration of the gastric mucosa. Vatistas *et al.* (1999) examined the gastric mucosa of two hundred and two thoroughbred racehorses endoscopically and found that 82% of horses had some degree of ulceration, and 39% showed clinical signs consistent with gastric ulceration. There is, therefore, a relatively high incidence of gastric mucosal disease in Throughbreds involved in active racing, which may be a reflection of the dietary conditions under which these animals are kept. Although ulceration is a rare cause of clinical cases of colic, it may produce more subtle effects on the welfare of horses, such as loss of condition and weight loss. There is some evidence, as well, that the pain of gastric ulceration is one of the precipitatory factors in the development of oral based stereotypies. This is because the appropriate behavioural responses needed to reduce or avoid the pain are inhibited by the animal's environment (Mason 1991). Indeed, the sub-therapeutic use of the antibiotic virginiamycin has been found to reduce the incidence of oral-based stereotypies in horses on high grain diets (Johnson *et al.*, 1998).

Stabling of horses can also cause problems through their enforced immobility. In a natural environment the horse would be constantly moving throughout the day and selectively grazing. Stabled competition horses spend a variable proportion of each day immobile, often followed by periods of challenging athletic activity. This sudden contrast in muscular activity can lead to problems such as rhabdomyelitis, or 'azoturia', where muscles fibres become damaged and painful (McLean 1973). MacLeay *et al.* (1999), in a study on thoroughbred racehorses, found that rhabdomyolysis is associated with diets high in soluble carbohydrate, and exercise below racing speeds. The frequency of the disease was found to be reduced by including fat and adequate, rather than excessive, carbohydrate as the energy source, and implementing an exercise programme aimed to reduce stress and excitability.

Stabling can also have an impact on the natural health of the equine foot (see Figure 1). The horny laminae that make up the outer part of the structure of the hoof wall allow moisture to pass in and out from the sensitive laminae. When the hoof wall remains in either damp conditions, or very dry conditions, for prolonged periods, the moisture content of the environment can cause changes in the integrity of the horny laminae. In damp conditions, the laminae can become soft and the wall becomes a less effective barrier to infectious agents, and in dry conditions the wall can become hard and brittle, which can lead to cracking, and also prevents the natural expansion and contraction of the foot that occurs as the horse puts weight on it. Inadequate hygiene in stables, causing damp conditions for the foot is probably now much less of a common problem than those situations where horses spend a majority of their time with their feet in very absorbent materials, such as shavings or bedding materials made from hemp. In combination with other factors, notably genetic effects on hoof structure, and shoeing practices, drying of laminae can be a factor in the development of a range of painful conditions of the feet. Where horses are stabled on absorbent bedding materials for the majority of the day, problems such as quarter cracks and sand-cracks are more likely to result in loss of integrity of the hoof wall (Richardson 1994). It seems likely that the pain of laminitis is worse where the hoof wall is brittle and inflexible to expansion.

2.2. GRAZING

Although grazing is obviously the 'natural' method of feeding for horses, and the method which best matches their motivational needs (see Chapter 1), there are a range of problems which occur remarkably regularly in horses in the UK that are related to grazing.

The most common of these is undoubtedly laminitis, which has also been called 'founder' or 'sinking syndrome'. This painful condition is caused by an inflammatory response in the sensitive laminae in the hoof, is not only responsible for a significant loss of athletic function amongst domestic horses (Cripps & Eustace 1999), it is also a significant cause of euthanasia. In a study carried out in France between 1986 and 1998 on 1040 horses which were examined post mortem to determine the cause of death, 13% of the deaths were due to locomotor disorders, and 8.2% of these were due to laminitis (Collober et al., 2001). Laminitis can arise for a number of reasons, such as secondary to Cushing's disease, hypothyroidism or trauma to the feet, but the most common pathogenesis is through the production of endotoxins from Gram negative bacteria in the colon, which proliferate where there is an excess of carbohydrate in the diet that is not digested in the small intestine. Laminitis has been recorded in Przewalski horses kept in a semi reserve (Budras et al., 2001), and hence can be regarded as a disease that occurs through management factors, rather than as a consequence of genetic changes that have occurred through the process of domestication. Indeed, it is those horses and ponies which are of the most ancient and hardy 'type', that are most likely to suffer from this condition, as they are able to make best use of the water soluble carbohydrate available in their diet. Carbohydrate induced laminitis mostly arises because the pasture that horses are kept on is generally of much superior quality in terms of digestible energy than the highly fibrous grasses that they evolved to survive on, and often domestic horses are given supplementary feeds in addition to this.

It is the same types of animals that are prone to laminitis that are also most likely to develop hyperlipidaemia, a condition which arises when animals that are overweight have a sudden reduction in energy intake, and their fat reserves become mobilised to a degree that causes damage to the liver. Sudden changes in food availability in ponies, for example, can lead to hyperlipidaemia (*e.g.* Jeffcott & Field 1985).

Another problem that affects horses that are grazed on unmanaged pastures is the risk of ragwort poisoning (Leyland 1985). The consequences of consuming ragwort even when intake is gradual over a prolonged period of time are very serious for horses, and result in liver disease, and ultimately liver failure (Ford 1973). Large worm burdens and associated damage to the intestinal mucosa and mesenteric arteries, are also more of a serious risk for horses at pasture, where infection can be picked up whilst grazing (*e.g.* Kaplan & Little 2000). A study investigating the prevalence of tapeworms, bots and nematodes in an abattoir population found that 55% had Gasterophilus intestinalis, and 48% Anoplocephala perfoliata, although numbers of nematodes were lower in horses treated with ivermectin, and tapeworms were lower where pyrantel had been used (Lyon *et al.*, 2000).

Traumatic injury, such as fencing injuries or injuries from intraspecific aggression are also more likely to occur whilst grazing than when stabled. The risk of injuries from features of the environment can be reduced by ensuring that fences and water troughs are safe for horses, and injuries from other horses are minimised by keeping horses within social groups that are as stable as practically possible (McGreevy 1996).

Myoglobinuria (otherwise known as 'azoturia' or 'tying' up) is a condition usually associated with stabled horses, which results in painful and damaged muscles and the passing of myoglobin in the urine making it a red colour. Atypical myoglobinuria is a rare form of myoblobinuria that occurs in grazing horses, which is often fatal because of widespread skeletal and cardiac muscle degeneration. The aetiology of this condition is unknown, although factors such as herbicide toxicity have been suggested (Whitwell *et al.*, 1988).

2.3. BREEDING

The mare is seasonally polyoestrus with a breeding season from May to October in the Northern Hemisphere, outside of which time the mare is anovulatory (Hughes *et al.*, 1975). In the Thoroughbred racing world, horses are aged as from the 1st of January each year, and hence for each age group, those born earlier in the year have an advantage in terms of maturity than those born later in the year (Langlois & Blouin 1996). A study by Moritsu *et al.* (1998) in Japan, for example, found significant differences in racing performance in horses born in January or February to those born in May, June or July, at four and five years of age. The effect of maturity on performance has led to the widespread use of hormonal preparations that induce or synchronise oestrus, or which establish normal cycling earlier in the year to ensure earlier foals (Ulmer 1985). Infusion of gonadotrophin-releasing hormone (GnRH), for example, is used to induce ovulation and fertile oestrus in mares during seasonal anoestrus (Hyland 1993). The use of these preparations changes the normal cycles of reproductive activity in the mare, and can have effects on behaviour and fertility, purely because of an arbitrary ageing date for foals that are born. In addition, valuable mares involved in competition have embryos trans-ferred to carrier mares to ensure that they can continue competing (*e.g.* Woods & Steiner 1986), with resultant changes in their reproductive hormone balance. The process of 'in hand mating' (where the stallion is led to the restrained mare, rather than the more natural process described in Chapter 1) is also highly unnatural for both mare and stallion (McDonnell 2000), and is likely to cause stress, especially for inexperienced animals. The stress associated with forced matings, where none of the normal pre-coital behaviours are possible, is one possible explanation for the reduced fertility in domestic horses to those, which are free ranging. This is because ACTH and cortisol have suppressing effects on the hypothalamo-pituitary gonadal axis at the level of the hypothalamus, pituitary and target gland levels (Dobson & Smith 2000).

Apart from the welfare issues surrounding the manipulation of mating, foaling can also be a cause for injury and fatality in horses. In a study conducted in France, for example, which investigated the causes of mortality in horses from claims to insurance companies, foaling was found to be the highest cause of mortality claims (24%), greater than colic (21%) and locomotor diseases (21%) (Leblond *et al.*, 2000). Pain can also be a cause of foal rejection by dams, as painful retention or passing of the placenta by the mare can be associated with the foal, resulting in aggression or avoidance, especially in inexperienced mares (Houpt & Feldman 1993; Juarbe-Diaz *et al.*, 1998). Similarly, udder pain, for example from mastitis, can be a direct stimulus for foal-directed aggression. Pain can also inhibit sexual behav-iour in stallions. This can be direct avoidance of pain in the penis, joints, feet or back (Houpt & Lein 1980), or a conditioned avoidance to a previous experience of pain in that context. Stallions fitted with a stallion ring to inhibit sexual behaviour when in training, or those which are punished for masturbation can develop a con-ditioned fear response that results in permanent impotence (Houpt & Lein 1980).

2.4. TRANSPORTATION

Transportation of horses, for competition or slaughter, is obviously a stressful psy-chological experience for horses, and in addition can be detrimental to their physical welfare. Horses travelling for a 24 hour journey, for example, were found to have a 6% loss of body weight, and also showed changes in muscle metabolism, stress indices, dehydration and immune parameters which increased with the duration of transportation, but which mostly returned to within normal ranges by 24 hours after the end of the journey (Stull & Rodiek 2000). The effect of transport on welfare is discussed in detail in Chapter 6. However, as well as the direct physiological effects of stress and dehydration on the horse, other aspects of transportation can lead to clinical problems that are detrimental to their welfare. The artificial position in which horses' heads are restrained during transportation, for example, is of concern because it prevents them from being able to brace themselves against changes of speed. In addition, an elevated head position over a prolonged period of time also leads to an accumulation of secretions and a significant increase in bacterial numbers, particularly Pasturella and Streptococcus species (Raidal *et al.*, 1995), which may well contribute to the incidence of respiratory disease subsequent to transportation. Transportation also generally leads to problems of infectious agents being spread by mixing of horses at their destination, and hence horses travelling to competitive events will be exposed to novel pathogens at times when they are stressed and immunocompromised. The stress of transport has been shown to have a deleterious effect on the phagocytic abilities of pulmonary macrophages, as does the stress of training or strenuous exercise, which may explain the increased incidence of pleuropneumonia in animals transported for competition (Sweeney *et al.*, 1985). Increased incidences of respiratory disease, for example, coincide with the seasonal influx of yearlings into racing yards (Burrell 1985). This is as true for horses engaging in other forms of competitive activity as it is for race horses (Australian Senate Select Committee 1991). Apart from being a factor in the development of respiratory disease (*e.g.* Hobo *et al.*, 1997), the stress of transportation has also been reported to be important in the development of colitis (Ecke & Hodgson 1996), rhabdomyolysis (*e.g.* Ito *et al.*, 1992) and hyperlipaemia (*e.g.* Jeffcott & Field 1985).

Physical injuries from trauma are a further risk to horses during transportation. The degree and nature of injury depends on the type of vehicle, direction of travel, skill of driver and stocking density (*e.g.* Collins *et al.*, 2000), and is discussed in more detail in Chapter 6.

3. Riding Related Problems

The fact that some horses are used for a range of strenuous and challenging competitive activities obviously predisposes these individuals to the likelihood of having more injuries than horses which are not ridden or competed. As would be expected, the major risk factor for lameness is use in any form of competitive activity, from racing to showing (Ross *et al.*, 1998). In this section some of the major considerations in terms of riding horses generally, and the specific challenges to welfare that are unique to certain competitive disciplines will be examined.

3.1. PAIN

The occurrence of pain in any species is a welfare concern. Obvious signs of painful lesions, such as lameness, occur more frequently in horses used for particular kinds of activities, as outlined later in this chapter. In addition, there is also a range of chronic conditions causing pain or discomfort to horses, which are poorly recognised by owners and carers of horses, and are sometimes even unknowingly inflicted by them. As well as frequently leading to acute disease, chronic pain or discomfort can also cause the horse to develop a behavioural response designed to avoid the painful stimulus. This is particularly a feature of the horse, being a social prey species that has evolved for grassland survival, it has a morphology, sensory system and behavioural repertoire designed for rapid flight from perceived dangers (Goodwin 1999). It is adaptive for the horse to take avoidance action from anything that it perceives as aversive. It is also evolutionarily adaptive for a horse to learn very rapidly about stimuli that are predictive of an aversive event. Flight is the main defence reaction (Bolles 1970) in the horse, and is elicited more easily by some stimuli than others (Seligman 1970). Equines are likely to learn avoidance responses very rapidly, with both fewer exposures and less intense exposures to the predictive stimuli, than domestic 'predator species', such as the dog. Horses will also learn avoidance responses faster than approach responses, because in evolutionary terms survival is more dependent on the former in an open grassland environment. Although flight is the natural response of the horse to aversive events, where this is not possible, defensive aggression is used to keep the 'threat' away. The occurrence of pain in horses is, therefore, an important factor in individual animals developing avoidance or defensive behaviours in order to remove itself from the source of pain. Recognition of this is particularly pertinent as the particular demands made of modern day competition horses make the occurrence of pain the single most likely cause of problem avoidance and aggression behaviours in the domestic horse population. The particular behaviour shown by an individual animal will vary with the nature of the painful stimulus, the situation in which the stimulus is present, and whether the behaviour occurs in response to the stimulus itself, or to an event that predicts the painful stimulus.

Pain perception in horses. The anatomy and physiology of both the peripheral and central nervous system of the horse strongly suggests that the horse is able to not only feel and respond to pain (Holliday 1972; Swenson 1993) but also has an emotional response to pain (Guthrie 1980). Horses also display a behavioural response to pain, as occurs with lameness, which alters with the use of analgesics (Gibson & Paterson 1985). They are generally very sensitive to touch and pain, although this sensitivity does vary with breed and type. The skin and viscera of the horse contain specific nociceptors, which respond to the local release of 'pain mediators' such as prostaglandins and bradykinins, released in response to tissue damage caused by heat, abrasion, pressure, or irritants (Iggo 1993). Firing of nociceptors increases with the severity of the stimulus applied. Six different types of receptors and a large number of free nerve endings detect other sensory information about the specificity of the stimulus. Information from these receptors travels to the spinal cord through large myelinated (A-beta), small myelinated (A-delta), and small unmyelinated (C) fibres, and then up the spinal cord via the six major spinal tracts. Some of these tracts are phylogenetically older, and slower, and these synapse in the reticular formation in the brain stem, especially the periaqueductal grey. The phylogenetically newer tracts are faster, and synapse in the ventrobasal complex of the thalamus and mid-brain (Gray 1987).

The pathways leading to the brain stem and thalamus increase the reactivity of the animal, causing a physiological stress response. They also connect to the limbic system, which activates the sympathetic nervous system, controls the endocrine responses, organises behavioural processes such as escape or aggression and instigates the development of the emotional component. The conscious perception of pain, including the strength and position of the stimulus, occurs in the parietal lobe of the cerebral cortex, which is well developed with multiple folds in equines. The perception of pain in this area is controlled by a 'gate-control' system (Melzack & Wall 1988). The final input depends on the characteristics (*e.g.* pressure, heat) and intensity of sensory input, the speed at which the signals travel up the spinal cord, and inhibitory effects from centres involved in fear responses, such as the amygdala, and higher centres (Gray 1987). The area of the brain that has control over the execution of a flight response, or defensive attack, is the periaqueductal grey of the midbrain. Tonic inhibition of output from this centre comes from the ventro-medial nucleus of the hypothalamus. This inhibitory pathway is itself subject to inhibition from the amygdala during a fear response, or is intensified by input from the septo-hippocampal system. Whether there is a behavioural response to a painful stimulus depends on the balance of inputs into the inhibitory pathway. The type of behavioural response most appropriate to the stimulus is controlled by the midline nuclei of the hypothalamus (Gray 1987).

Painful stimuli in the horse therefore give rise to not only painful sensory experience, but also a behavioural response and an emotional component. The fast tract pathways mean that reactions to painful stimuli are very rapid, but are generally predictable in nature to each given stimulus.

Anticipation of pain. At first examination it seems that fear and pain can provoke similar behavioural responses in the horse. In both cases negative effective components elicit physiological and behavioural responses which are aimed at avoiding or minimising the conditions which produce them. However, the processes for the elicitation of these behaviours are different for each motivation, although in many circumstances the two occur simultaneously. The direct response to a painful event may cause a different behaviour pattern from the response to a stimulus conditioned to a painful event. The direct response to pain occurs through the pain pathways described previously, but the conditioned response is a fear response occurring due to the prediction of a painful event (Azrin 1967). For example, it is common for aggression to be directed at nearby individuals in response to a direct pain stimulus, such as intramuscular injection, but such a response is less likely in response to a conditioned pain response, such as the arrival of the veterinary surgeon. The behavioural pattern shown in the latter case is a fear response due to the anticipation of pain, and generally has an avoidance pattern. In addition, analgesia will reduce the response of an individual to a painful stimulus, but not reduce the fear response to a stimulus conditioned to a painful event. Anxiolytic therapy has the opposite effect, whereby the conditioned response is decreased to a greater degree than the pain response (Gray 1987).

These different responses arise because the conditioned fear response is controlled by a completely different system in the brain to that which controls the direct response to pain. Indeed, the presence of a conditioned fear response will actually interfere with the response to a novel painful stimulus (Fanselow & Baaches 1982), through the release of endogenous opiates. This response causes animals, and humans, apparently not to feel pain during periods of emotional arousal. This is important in the recognition of pain, because a behavioural response may be occurring in individuals because there is a directly painful stimulus, or because the horse is avoiding a situation that it has learnt through previous experience causes pain.

The identification of pain in horses is very subjective, and although experienced veterinary surgeons in equine practice often have some ability to assess the need for analgesia instinctively, there is huge variation in the consistency of this assessment. The level of pain experienced by their horses is also something that is often misinterpreted by owners and carers. Animals that experience chronic pain and exhaustion, such as in prolonged bouts of colic, generally have a loose lip but clenched masseter or cheek muscles (Houpt & Fregin 1992). In acute pain, the attention of the animal is directed towards the affected part of body, by turning the head, or moving the ears towards the source of the pain. Where vocalisation is present it is generally a squeal in somatic pain, and a grunt where the pain is visceral. The facial expression is described as that of 'concern', generally with a fixed stare, with the eyes less mobile than usual in the orbit and the eyelids puckered, and the ears slightly back and fixed in position. Most horses have slightly dilated nostrils (Fraser 1992). This is different from a fear response where generally the eye is very mobile in the orbit, the ears are mobile, the cheek muscles relaxed but the lower lip tense. Where a fear response is present it will override the expression of pain, so the possibility of underlying pain being present should not be overlooked where there is also a conditioned fear response. Where pain occurs in the musculoskeletal system, there is often tension in associated muscle masses in order to guard the damaged tissue from further injury. There is also often the adoption of postures that help to minimise the incidence of pain, such as the characteristic posture of horses with laminitis. Interpreting the degree to which an individual horse shows pain is also complicated by the fact that there is an enormous variation in the degree to which individuals respond to an equivalent painful stimulus. As in humans, some individuals are more stoical and tolerant of pain or discomfort than others, which tends to make pain assessment in practice very subjective (Chambers et al., 1993).

Behavioural changes may occur in an individual horse due to direct acute pain, chronic or sub-clinical pain, or through conditioned responses from the previous experience of pain. Veterinary examination is therefore essential before 'behavioural problems' in horses are treated, as active sources of pain must be identified and treated before the horse can be taught anything different about the predictive stimuli for that pain. It is important to consider that the painful response may not be present in the standing horse, and the history of the problem is likely to give clues as to the context in which the painful response is present. Muscular injuries, for example, may only be significant during particular forms of activity, and pain associated with saddlery or inexperienced riding will only be present during tacking up or ridden exercise.

The commonest sources of pain involved in the development of avoidance or aggression responses in the horse are those involving muscular injury, especially to the back, and injuries to the mouth, such as occurs due to sharp spurs on molar teeth and poorly fitted or inappropriate bit use. Back pain in horses is usually due to soft tissue injuries, particularly of the muscles and ligaments of the thoracolumbar spine (Jeffcott 1979) as subluxation of vertebrae is physically highly unlikely. Correct fitting and use of saddles is a very important aspect of equine welfare, and poorly fitting saddles a very common cause of chronic back pain and poor performance in horses (Harman 1995). A study conducted by Jeffcott *et al.* (1999) investigated the use of a pressure sensitive mat to identify areas of increased pressure on a horse's back under a saddle. The mat was used in 10 horses to investigate saddle fit, and evaluate the accuracy of the technology. A linear relationship was found between the total force (weight) exerted and the pressure measured beneath the saddle on a wooden horse and a standing live horse. It was also found that the pattern of pressure on the horses' backs changed with pace – *i.e.* there were different centres of pressure at walk, sitting trot, rising trot and canter, which is an important consideration when saddles are fitted on horses at rest and walk only.

The effect of girth tension on the athletic performance of racehorses in Australia was examined by exercising horses on a treadmill using a randomised block design, with a girth tension of approximately 5, 10, 15 or 20 kg (Bowers & Slocombe 1999). This study found that with each increase of girth tension during inspiration and expiration at rest, there was a reduction in the distance that the horse travelled before tiring, and the time spent exercising. It was found that girth tensions of less then 10 kg were found to be optimal for performance, although tensions greater then this was typical for racehorses in Australia, as well as being very variable between individual horses and grooms (Bowers & Slocombe 2000). There is no evidence, however, whether the decrease in performance is due to discomfort/pain, or whether there is a decreased inspiratory air flow or volume.

The other major cause of pain in ridden horses arises from differing riding styles and abilities. Rough, poorly balanced or uneducated riders will inadvertently cause pain, by, for example putting more weight on one side than the other, using excessive force to move the horse forward, or catching the horse in the mouth with the bit due to sudden hand movements. The effect of poor riding on horses is exacerbated when jumping, where imbalance can cause the rider to land heavily on the horse's back, or pull back on the reins to balance themselves on landing. Pain in such circumstances commonly causes the horse to learn to avoid jumping, for example by refusing. Such horses are often encouraged to jump through punishment for stopping, making the entire experience a very unpleasant one for the horse. Some horses will avoid entering jumping arenas to avoid the discomfort and punishment associated with jumping.

Foot pain, other musculoskeletal injuries, eye conditions, aural plaques and ear mites, focal bruising and veterinary procedures are other common causes of avoidance or defensive aggression responses. As discussed, any of these painful foci can cause a direct response, or can lead to a conditioned fear response to an associated stimulus, and the behaviour seen is very often different in each case. For example, a dressage horse with myositis in the semitendinosus muscle of the thigh may exhibit a shortened stride, resistance to engagement or aggression in response to pain when worked (Denoix & Pailloux 1996). A conditioned fear response to such an injury may be associated with the arena, and cause an avoidance response to entering on the next training session, even if the lesion has resolved.

The higher incidence of pain related changes in behaviour is, therefore, partly related to the particular sensory and behavioural adaptations of the horse as a prey species, and partly to the fact that greater athletic demands are made on equines than other companion animals. It is essential, therefore, that the presence of painful loci is considered in the welfare of competition horses, and the identification and treatment of underlying sources of pain by a veterinary surgeon, and physiotherapist where indicated (Bromiley 1993), must be achieved before other forms of modification of behaviour can be successfully accomplished.

3.2. HEADSHAKING

A typical headshaker is a horse which displays recurrent, intermittent, sudden and apparently involuntary bouts of head tossing which may be so extreme as to throw the horse and rider off balance. Sneezing and snorting are frequently present, accompanied by attempts by the horse to rub its nose on the ground, a foreleg or nearby objects. (Mair & Lane 1990; Madigan 1996). Although the condition can occur at rest, more cases are obvious during exercise, especially at trot (Cook 1979a; Lane & Mair 1987; Madigan et al., 1995). It can, however, be seen at any pace.

Cook (1979a, b, 1980a, b) records 58 disorders which may show headshaking as a presenting sign but establishing a diagnosis either ante- or post-mortem is often very difficult. For example, Lane and Mair (1987), working in a referral centre, could identify a potential cause in 11% of their cases and found correction alleviated the problem in only 2%. Whilst a seasonal pattern of symptoms may suggest an allergic rhinitis or condition linked to photoperiod (Lane & Mair 1987; Mair & Lane 1990; Madigan et al., 1995), it must be remembered that many other factors also correlate with this time of year. This includes an increased tendency to exercise the horse and a higher aerial dust burden, both of which may be implicated in irritation leading to headshaking. The occurrence of certain specific behavioural features may suggest irritation of a specific area, for example, horses which shake their head horizontally, either in addition to or instead of the classical vertical movement, or rub their ears may be more likely to have a pain focus around the ears, such as might occur with ear mite infestation (Mair & Lane 1990; Mayhew 1992). 'Flipping' the nose (a characteristic twitching movement), sneezing, snorting, 'clamping' the nostrils as if to close them and attempts by the horse to rub its face on a foreleg are all reported to be responses to naso-facial irritation or pain (Madigan & Bell 1998). Since many cases remain without a definitive diagnosis or effective treatment, they may be labelled as a form of stereotypic or behavioural disorder although a physical cause is suspected in the majority of cases (Mair & Lane 1990). Stereotypic behaviours are repetitive, relatively invariant behaviours with no obvious function (Mason 1991) and include problems such as crib-biting, weaving, windsucking and boxwalking (Mills & Nankervis 1999). It is arguable whether or not any headshakers belong in this category, although a recent report (Cooper et *al.*, 2000) suggests a stereotypic form of nodding may occur in the horse, which could theoretically be confused with headshaking. This condition is however amenable to treatment through changes in management. Thus a true headshaker, should be recognised as an individual in pain although the exact area affected may vary between individuals. Some horses headshake in response to specific stimuli, such as pressure on the bit, or change of rider weight, and in these animals, such responses are likely to be behavioural responses to discomfort or pain. Recent work (Mills *et al.*, 2001a) has sought to analyse the signs in order to identify clusters of symptoms suggestive of the different areas affected, with the actual headshaking being the most obvious and linking sign in response to the head pain. This might offer future hope for more precise treatments.

Because headshaking has a heterogeneous aetiology and a diagnosis rarely made, treatment also remains problematic despite recent advances (Newton *et al.*, 2000). Mills *et al.* (2001b) in a survey of 245 owners of headshakers, found about three-quarters of owners had consulted a veterinary surgeon about their horse's condition, but traditional veterinary therapy was used in only 51.8% of cases. The number and proportion of owners using each treatment are given in Table 1. The most popular feed supplements were herbals, such as Echinacea, which is thought to boost the immune system and respiratory function (Bauer & Wagner 1991). The 'other methods' included: inhalation of friar's balsam, vinegar and oil rubbed on the muzzle before exercise, sun-block, a leather fringe on horse's nose-band, feeding only grass and hay, keeping the horse in an open barn, riding at dawn and moving the horse to another location. However, recent work by Newton *et al.* (2000) suggest the potential for much better success with a combination of cyproheptadine and carbamazepine. The most effective non-veterinary management aid was the nose net, with more than a quarter of all users reportedly achieving complete success. This has traditionally been thought to be effective by acting as a filter but the recent success of an alternative 'half-net', with a large mesh (Net Relief Equilibrium Products, UK) suggests that an alternative mechanism may be involved in their success. The most likely explanation relates to a reduction in turbulent airflow

Table 1. Reported success of treatments for headshaking used by owners (from Mills *et al.* in press).

Treatment	Number of horses treated in this way	Complete success	Partial success	No success
Traditional vet. treatment	**127 (51.8)**	7 (6.3)	28 (22.0)	93 (73.2)
Back specialist	**50 (20.4)**	0 (0)	4 (8)	46 (92)
Homeopathy	**93 (38.0)**	6 (6.4)	29 (31.2)	58 (62.4)
Other alternative therapies	**39 (15.9)**	2 (5.1)	5 (12.8)	32 (82.1)
Nost veil/net	**179 (73.1)**	48 (26.8)	61 (34.1)	70 (39.1)
Ear net	**82 (33.5)**	3 (3.7)	24 (29.3)	55 (67.1)
Face net	**51 (20.8)**	4 (7.8)	21 (41.2)	26 (51.0)
Feed supplement	**105 (42.9)**	5 (4.8)	32 (30.5)	68 (64.8)
Other methods	**91 (37.1)**	10 (11.0)	39 (42.9)	42 (46.1)

Figures in bold parentheses represent % of total population surveyed ($n = 2245$).

through the nose as a result of the mesh. It is thought that turbulent airflow results in stimulation of sensitised nerves within the nose.

This hypothesis also explains why tracheostomy or chemical ablation of the posterior ethmoidal branch of the trigeminal nerve is successful in some cases (Newton *et al.*, 2001), especially when resection of the infra-orbital branch of the nerve which relays sensation from more superficial structures has failed. It may also explain why the problem is more common at the trot and possibly in dressage horses, since the angle of the head during these exercises may encourage a more turbulent air flow pattern. Interestingly, the condition appears to be very rarely seen in race horses, which usually exercise with their necks outstretched causing minimal resistance to airflow.

In conclusion, headshaking remains a problematic condition affecting the ridden horse, but there have recently been significant improvements in our understanding of the condition, leading to improved prospects for treatment which has traditionally been very poor. The term 'headshaker' should be regarded as a description of the presenting problem and not the definitive diagnosis. It is one indication that the horse's welfare is compromised and that action is required to rectify the situation.

4. Welfare Concerns Associated with Specific Competitive Disciplines

In this section a few of the competitive activities for which horses are used will be reviewed to highlight the specific welfare concerns that are unique to each. The sections are not exhaustive – obviously there is a huge range of potential diseases and injuries that can occur to horses in any of these disciplines – but the aim is to encourage the reader to reflect on the likely impact of different activities on the physical welfare of horses that are used for them. Each type of competitive activity involves a unique challenge in terms of muscular and athletic abilities in the horses, which predispose those horses to particular types of injury or disease. As well as the specific demands of each type of activity, the conformation that is selected for in animals destined for that discipline is an important factor in the development of particular injuries or diseases. Estimates of heritability have been compared, for example, against conformation type, performance ability and bone disorders in competition horses (Willms *et al.*, 1999). Concern about the welfare of horses involved in the racing industry involves more public debate than other disciplines, because of the popularity of, and controversy over, the more well known races, such as the Grand National in the UK, and publicity over the 'wastage' of thoroughbreds from the industry. As well as the dangers of injury involved in racing itself, other management practices for horses used in this industry are of concern with regard to their welfare. The significant problems specific to the racing industry are discussed in Chapter 8.

Endurance. Because the competitive sport of endurance riding in the UK was set up in careful consultation with the veterinary profession, veterinary examinations are integral to all the events that are held in this sport (Hall-Patch *et al.*, 1977), helping to ensure that the welfare of their horses is considered as a priority by competitors. These checks include a pre-ride examination which should prevent those horses which are unfit for the event from participating (Blake-Caddel 1992), mid way checks, and checks at the end of the ride to ensure that horses are not over exerted in the final stages of the race. This strict regime of regulated examinations helps to ensure that the welfare of horses involved in this discipline is not compromised through over exertion, dehydration or lameness, and has also spawned a number of scientific papers that examine the effect of prolonged exercise on various physiological and other biological parameters in the horse. A review by Fowler (1980) identified the major veterinary problems that arise during endurance trail rides, which were mainly related to hyperthermia and dehydration. Muscular disease, including those induced by the build up of lactic acid within the muscles due to anaerobic metabolism, laminitis, lameness, colic and heart disease were also recorded as less frequent causes of illness during endurance rides. Significant changes do occur in most parameters of plasma biochemistry indicative of a stress response in horses competing in an endurance ride, as compared to control samples taken before the ride, and these changes, as well as elevated heart rate, are still apparent 30 minutes after completion of the ride (Rose *et al.*, 1977). Within reason, these changes do not necessarily indicate that the animals are suffering – dehydration, and low levels of sodium, potassium, magnesium and chloride salts are present in all horses at the middle check point of a 100 mile ride, and low electrolyte levels are also present at the end of the ride (Carlson & Mansmann 1974). These changes will normally occur with prolonged physical exertion, and will return to normal levels with rest and re-hydration, although recovery is speeded up with the use of supplemental salt preparations. Lucke and Hall (1980) found that horses completing the 'Golden Horseshoe Ride', a 100 mile ride ride in the UK, were moderately dehydrated, and were apparently working aerobically using fat as a substrate for metabolism, but had no signs of post exercise ketosis. Biochemical signs of more severe dehydration, variations in acid-base balance, and a marked haematological stress response, are found in horses that are exhausted and unfit to continue (Carlson *et al.*, 1976), and generally such individuals are identified at veterinary checks and prevented from continuing further. In rides that run over several days, although horses do not experience progressively greater body weight losses through dehydration, they may well experience progressive electrolyte depletion (Schott *et al.*, 1996), and hence for these prolonged competitions, biochemical signs of exhaustion need to be monitored more carefully. Another study, investigating the cardiac recovery of competing endurance horses, found that an increase of four or more beats per minute in heart rate after trotting up during the midway check, indicated fatigue or pain. Poor cardiac recovery rates were particularly noticed in horses that were later stopped from the ride because of lameness, suggesting that the earlier decline in cardiac performance was linked to musculoskeletal pain (Trevillian *et al.*, 1997). Another factor that needs to be taken into consideration by veterinary examiners is the condition of horses prior to starting endurance – in a study carried out by Garlinghouse and Burrill (1999), for example, condition score had a significant effect on completion rate of a 160km ride, and the distance covered for horses that did not complete. Lawrence *et al.* (1992) also found that horses that completed a 150 mile race were more likely to have a lower condition score, and have less rump fat, particularly when compared with those horses eliminated from the race on metabolic criteria.

The importance of adequate training is illustrated in the study conducted by Hodgson and Rose (1987) which examined muscle biopsies of horses prior to training, at three months, six months and nine months of training, and again three months after the training ceased. The types of fibres present in the gluteal muscle changed over the period of training to include significantly more Type IIB fibres with a greater oxidative capacity to enable the animal to cope with prolonged exertion. In addition, there were significantly more capillaries leading to the muscle fibres, improving the oxygen and glucose supply to muscles to increase their aerobic capabilities, and an increase in activity of oxidative enzymes within the muscle fibres. Although the composition of fibres remained as for the training period during the three months after the training was ceased, the number of capillaries and activity of enzymes involved in the oxidative process had declined, reinforcing the importance of a continued level of training in horses expected to compete to avoid muscular and metabolic abnormalities. The most successful horses in endurance competitions have a higher proportion of slow twitch high oxidative fibres (Snow *et al.*, 1981).

Eventing. The sport of eventing creates quite unique challenges in terms of equine welfare, as horses competing in this sport need to display a high level of athletic ability involving different types of movement, and also be fit enough to sustain a prolonged period of high intensity activity (see Figure 2). The particular kinds of injuries that are received by event horses obviously relate to the three specific phases of the competition, although the cross country phase is the one where most injuries occur – lameness and traumatic injury to other parts of the body are obviously of greatest welfare concern here (Hertsch 1992). Although there has been pressure for the complexity of cross country challenges to be increased over the past ten years, the importance of the welfare of competitors and their horses does lead to a constant revision of the courses and maintenance of the ground by course designers to minimise the risk of traumatic injury.

Most of the studies of eventing horses have concentrated on the effects of dehydration and physiological changes of competing at a high intensity for a prolonged period and over several days (see Figure 2). These studies were particularly aimed at reducing the impact of high intensity exercise on horses competing abroad, where ambient temperatures and humidities are higher than that to which the animals were accustomed. The response of European horses to the hotter and more humid conditions in North America, for example, resulted in the development of a trial that ensured sufficient time for recovery and heat dissipation, and the use of water showers to help heat dissipation (Kohn *et al.*, 1995). Horses competing in Burghley Horse Trials in the UK, which is a 4 star three-day-event (the highest level of difficulty), also had physiological, metabolic and biochemical parameters

Figure 2. Thoroughbred horse jumping a cross-country fence during an event (courtesy of Natalie Waran).

measured. The overall conclusion from this study was that in a temperate climate, the speed and endurance phases of the event do not induce extreme metabolic and physiological changes, and the majority of horses that finish have a rapid recovery from their exertion (Marlin *et al.*, 1995). Another study found that losses of total body water, sodium and chloride increased with level of competition, and that there was little recovery from dehydration 30 minutes after completing the cross country phase, and some deficits were still present the next morning, when horses were undertaking the next phase of the competition (Ecker & Lindinger 1995). Measures of serum electrolytes taken from horses competing at three-day-event competitions for other studies also revealed dehydration, metabolic acidosis and a compensatory respiratory alkalosis after the speed and endurance phases (Rose *et al.*, 1980, Hinchcliff *et al.*, 1995). As with the endurance horses, the horses re-hydrated in the overnight period, but sodium, potassium, chloride and calcium deficits were still present after 14–18 hours of recovery (Hinchcliff *et al.*, 1995). During competition, lactate and glucose concentrations are also increased in the plasma, and an increased activity is found in lactate dehydrogenase and aspartate aminotransferase, indicating a leakage of enzymes from working muscles (Andrews *et al.*, 1995). Plasma lactate was found to be highest immediately after exercise, but declined significantly after 10 minutes (White *et al.*, 1995). Body weight and total water were significantly decreased compared to baseline levels, at all stages of an event, with signs of dehydration being most acute in the endurance phase of the event (Andrews *et al.*, 1994).

The evidence suggests, therefore, that although horses competing at the top level of competition may have metabolic changes associated with exertion and dehydration during an event, the rate of recovery of these animals is very rapid, and are unlikely to cause compromised welfare. In addition, veterinary examinations at events are such that animals unfit to continue in a competition are prevented from doing so. The major welfare concern for horses competing at events is, therefore, the risk of traumatic injury through exertion or accidents occurring particularly in the cross country phase.

Dressage. Many of the problems that affect dressage horses are related to stabling for prolonged periods, as discussed in an earlier section. However, in addition to the respiratory and gastrointestinal diseases that occur in common with other stabled competition horses (Lindner & Schoneseiffen 1996), dressage horses are also susceptible to specific musculoskeletal problems that occur due to the particular demands of the discipline (Marcella 1990). However, there are generally less tendon, bone and joint disease of the lower limbs in dressage horses when compared with horses involved in show jumping or eventing (Hertsch 1992). Understanding the kinds of injury to which horses involved in competitive dressage are prone is aided by the use of time motion analysis of horses competing at Grand Prix level. This showed that such horses work at slow paces for 88% of the test, and at extended or faster paces for only 10% of the time. Their heart rate varies between 62 and 141 beats per minute, which is within the aerobic range of the horse (Clayton 1990). Dressage horses, therefore, are unlikely to suffer from the conditions related to over exertion, such as heat exhaustion and cardiac or respiratory distress, or have problems relating to the build up of lactate after periods of anaerobic activity.

Problems associated with dressage often involve the particular gaits that are desirable in this discipline. Studies of horses that have been schooled to a high level in dressage have been found to have; a greater duration of each stride, a shorter suspension phase and stride length, greater flexion of the shoulder and hock, and greater and more prolonged hyperextension of the carpal and fetlock joints, than untrained horses. This is thought to be due to the greater degree of upward 'thrust' needed for each stride (Holmstrom *et al.*, 1995; Morales *et al.*, 1998a, b). Certain conformation types of horses are obviously better 'designed' to perform the required degree of collection than others. Long backed horses, such as thoroughbreds, have difficulty rounding the back, bringing the hind legs under the body, and achieving the resultant increase in hock flexion and weight bearing, than horses of a more compact type, such as Andalusians. A study of the specific conformation features in warmbloods in Germany (Back *et al.*, 1996) showed that horses with a more horizontal scapula and a vertical pelvis showed the greatest ability to flex the shoulder and elbow in the forelimbs and hip and stifle in the hindlimbs. This allows for a greater degree of upward movement, and longer stance duration, improving the elegance of paces in dressage, and the forelimb technique whilst show jumping. The same study, however, also showed that horses which had a greater degree of extension in the carpal and tarsal joints, showed extension of these joints whilst weight bearing, which would be likely to lead to injury and loss from competitive use of these individuals. Hence, selecting the correct type of horse with appropriate limb conformation is important for those engaging in competitive dressage. Animals with other types of conformation are likely to find the degree of joint flexion required for advanced dressage difficult to achieve, and be much more prone to injury.

There is an additional problem, however, with horses that are selected specifically for their conformation as dressage horses, particularly those of warmblood origin. These individuals are often large but have a relatively short trunk as a consequence of the more vertical angle of the pelvis. Such animals have a much greater incidence of the painful and debilitating condition of overriding dorsal spinal processes, or 'kissing spines syndrome'. This problem arises from a combination of the conformation of the types of horse preferred in dressage, particularly on continental Europe, and the necessity for such animals to display a high degree of collection whilst carrying a rider. The dorsal spines of the thoracolumbar region essentially rub on each other causing pain, causing affected horses to show reluctance when asked for collection, to being ridden at all, or even to being saddled or handled. Although surgical resection of the spinous processes is attempted in some cases, it is not always a successful procedure, with 70% judged as successful by owners and 30% not (Lauk & Kreling 1998).

Show Jumping. As with dressage horses, the conformation of jumping horses is very relevant to their ability and performance, as well as to the kinds of musculoskeletal problems to which they are susceptible (see Figure 3). In a test of conformation against performance, upper forelimb width, depth of girth, width of lower forelimb, knee length, shoulder blade depth, shoulder joint angle and front fetlock joint angle, were all highly correlated to performance in a Grand Prix jumping competition (Slade & Branscomb 1989). These features correlated poorly with success at either dressage or racing, suggesting that these particular features are uniquely important for success at jumping. As with dressage, this data suggests that horses with other types of conformation (as compared with the ideal conformation described above) would be more prone to injury from jumping. The commonest kinds of injuries in show jumping horses are musculoskeletal injuries, particularly in the lower limbs, the genesis of which can be traced to the centres of load during movement (Hertsch 1992). Such injuries particularly occur on landing after jumping, and are exacerbated by increased height of the jump and hardness of the ground. Hence a proportion of these may be avoidable if horses are not competed where conditions are unsuitable for their welfare, although pressure to compete in all conditions is often a factor that encourages competitors to risk the soundness of their horse.

In terms of physiological parameters for show jumping horses, significant changes in acid-base status was found in 17 horses examined at a show jumping competition (Aguilera-Tejero *et al.*, 1998), with a significant decrease in pCO_2, bicarbonate, base excess, chloride and calcium, and a slight decrease in pH. Horses also showed an increase in haemoglobin, sodium, potassium, albumin and lactate

Figure 3. Show-jumping imposes specific risks of injury related mainly to the landing phase (courtesy of Natalie Waran).

in this study. As with the other competitive activities where the duration of exertion is limited, the rate of recovery from such physiological changes in show jumpers is likely to be very rapid unless the degree of exhaustion or dehydration is extreme.

One of the main concerns with regard to the welfare of horses engaged in show jumping is related to the popularity of the sport for amateur riders. As mentioned earlier the effects of poor riding skills on the balance and comfort of the horse are compounded by jumping, and many horses jumped regularly in small local shows and gymkhanas find the experience uncomfortable or aversive for this reason.

Polo. Polo ponies also develop a range of problems that are unique to the particular activities that they are used for. Polo requires the horses to be athletic and keep up high speeds for prolonged periods of time, but also requires a high degree of agility for sudden turning and stopping. Blood samples taken from polo ponies after competitive activity showed an increase in lactic acid, protein, sodium, haemoglobin, packed cell volume, and pH, and a decrease in chloride calcium, carbon dioxide and bicarbonate, compared to when at rest (Craig *et al.*, 1985). The changes apparent in polo ponies, of high plasma lactic acid with alkalaemia, therefore, partly mirror those occurring in racehorses where there are shorter periods of intense exercise, but also partly resemble those of horses competing in eventing and endurance events, where activity is sustained over a longer period. This is consistent with the fact that polo ponies have short periods of intense activity intermittently during a game, which requires both stamina and periods of intense speed. The ponies do, therefore, develop problems related to high exertion similar to race horses, such as exercise induced pulmonary haemorrhage (Voynick & Sweeney 1986), and also to problems similar to those of endurance horses, such as dehydration and myopathies. The incidence of such problems seems to be less in polo ponies than either the racing or eventing/endurance horses, probably because the period of sustained effort is generally less prolonged. More problems in polo ponies are caused by traumatic injuries (Marcella 1991), followed by muscular injuries and tendonitis (Schatzmann & Klemm 1998). Although the incidence of traumatic injury, particularly severe injury, tends to be higher in polo ponies, especially to the head (*e.g.* Tong 1994) and limbs (*e.g.* Gerring and Davies 1982), the overall incidence of injury was found to be lower than other forms of competitive activity in one study in Germany (Schatzmann & Klemm 1998). A Malaysian study, however, that investigated the clinical and radiographic findings in 33 working ponies found that 14 had fetlock joint swellings, 9 had restricted joint flexion, 13 had pain on flexion and 13 had lameness after a flexion test. A total of 23 of the 33 also had changes visible on radiographic examination of the distal third metacarpal, first or second phalanx (Bashir 1990). The incidence of proximal suspensory desmitis has been reported as being higher in polo ponies and jumping horses than those used for other purposes, presumably because of the sudden sharp turning required by riders in these disciplines (Personett *et al.*, 1983). The differences in the conclusions of these individual studies may reflect different training and competitive practices in the different countries in which they were conducted, or differences in the profiles of ponies studied. There is evidence overall, however, that traumatic injury is a significant problem in polo ponies, and the incidence of musculoskeletal injury is comparable to that found in other disciplines.

5. Conclusion

Although the horse has generally adapted remarkably well to a human controlled environment, the limits to which adaptation can occur is restricted its the degree of flexibility in both behavioural and physical phenotype. Beyond these limits the welfare of the animal is compromised. For the modern day sport horse, this compromise occurs in terms of the manner in which it is fed, housed, exercised and competed, all of which have diverged significantly from the environment, diet and diurnal activity pattern of the horse in the 'natural' environment. This means that the onus is on the owners and carers of horses, to ensure that the physical and psychological tasks required of horses are within the natural range of their abilities. It is also imperative that we recognise the signs indicative of stress or distress in horses, so that situations where the welfare of the horse is compromised can be resolved and avoided in future.

6. References

Aguikera-Tejero, E., Bas, S., Estepa, J.C., Lopez, I., Mayer-Valor, R. and Rodriguez, M. (1998) *Acid-Base Balance after Exercise in Show Jumpers*. Conference on Equine Sports Medicine and Science, 24–26 April 1998, Cordoba, Spain. Wageningen, Netherlands, pp. 43–45.

Andrews, F.M., Ralston, S.L. Ralston, S.L., Sommardahl, C.S., Maykuth, P.L., Green, E.M., White, S.L., Williamson, L.H., Holmes, C.A. and Geiser, D.R. (1994) Weight, water, and cation losses in horses competing in a three-day event. *J. American Veterinary Medical Association* **205, 5**, 721–724.

Andrews, F.M., White, S.L., Williamson, L.H., Maykuth, P.L., Geiser, D.R., Green, E.M., Ralston, S.L. and Mannsman, R.A. (1995) Effects of shortening the steeplechase phase (Phase B) of a three-day-event. *Equine Veterinary J.* **20, suppl.**, 64–72.

Australian Senate Select Committee on Animal Welfare (1991) Equine welfare in competitive events other than racing. Recommendations and major conclusions. *Australian Equine Veterinarian* **9, 4**, 137–140.

Azrin, N.H. (1967) Pain and aggression. *Psychology Today* **1, 1**, 26–33.

Back, W., Schamhardt, H.C. and Barnveld, A. (1996) The influence of conformation on fore and hind limb kinematics of the trotting Dutch Warmblood horse. *Pferdeheilkunde* **12, 4**, 647–650.

Bashir, A. (1990) Clinical and radiographic examinations of fore fetlock and pastern joints in working polo ponies. *J. Veterinar Malaysia* **2, 2**, 111–116.

Blake-Caddel, M. (1992) Pre-ride examination of horses entered for endurance competition. *Proceedings of the 37th Annual Convention of the American Association of Equine Practitioners*. San Francisco, California, December 1–4, 1991. American Association of Equine Practitioners, Lexington, USA, pp. 807–809.

Bolles, R.C. (1970) Species specific defence reactions and avoidance learning. *Psychological Review* **77**, 32–48.

Bowers, J.R. and Slocombe, R.F. (1999) Influence of girth strap tensions on athletic performance of racehorses. *Equine Veterinary J.* **30, suppl.**, 52–56.

Bowers, J.R. and Slocombe, R.F. (2000) Tensions used on girths on Thoroughbred racehorses. *Australian Veterinary J.* **78, 8**, 567–569.

Bromiley, M. (1993) *Equine Injury, Therapy and Rehabilitation*, Second edition. Blackwell Science Ltd, UK.

Budras, K.D., Scheibe, K., Patan, B., Streich, W.J. and Kim, K. (2001) Laminitis in Przewalski horses kept in a semi-reserve. *J. of Veterinary Science* **2, 1**, 1–7.

Burrell, M.H. (1985) Endoscopic and virological observations on respiratory-disease in a group of young thoroughbred horses in training. *Equine Veterinary J.* **17, 2**, 99–103.

Carlson, G.P., Ocen, P.O. and Harrold, D. (1976) Clinicopathologic alterations in normal and exhausted endurance horses. *Theriogenology* **6, 2/3**, 93–104.

Carlson, G.P. and Mansmann, R.A. (1974) Serum electrolytes and plasma protein alterations in horses used in endurance rides. *Journal of the American Veterinary Medical Association* **165, 3**, 262–264.

Chambers, J.P., Livingston , A., Waterman, A.E. and Goodship A.E. (1993) Analgesic effects of detomidine in Thoroughbred horses with chronic tendon injury. *Research in Veterinary Science* **54, 1**, 52–56 .

Clayton, H.M. (1990) Time motion analysis in equestrian sports: the Grand Prix dressage test. *Proceedings of the Annual Convention of the American Association of Equine Practitioners* **35**, 367–373.

Collins, M.N., Friend, T.H., Jousan, F.D. and Chen, S.C. (2000) Effects of density on displacement, falls, injuries and orientation during horse transportation. *Applied Animal Behaviour Science* **67, 3**, 169–179.

Collober, C., Foucher, N. and Moussu, C. (2001) Causes of mortality in adult horses. *Equ'Idee* **40**, 34–36.

Cook, W.R. (1979a) Headshaking in horses: part 1. *Equine Practice* **1**, 9–17.

Cook, W.R. (1979b) Headshaking in horses: part 2. History and management. *Equine Practice* **1**, 36–39.

Cook, W.R. (1980a) Headshaking in horses: part 3. Diagnostic tests. *Equine Practice* **2**, 31–40 .

Cook, W.R. (1980b) Headshaking in horses: part 4. Special diagnostic procedures. *Equine Practice* **2**, 7–15.

Cooper, J.J., McDonald, L. and Mills, D.S. (2000) The effect of increasing visual horizons on stereo-typic weaving: implications for the social housing of stabled horses. *Applied Animal Behaviour Science* **69**, 67–83.

Craig, L., Hintz, H.F., Soderholm, L.V., Shaw, K.L. and Schryver, H.F. (1985) Changes in blood constituents accompanying exercise in polo horses. *Cornell Veterinarian* **75**, **2**, 297–302.

Cripps, P.J. and Eustace, R.A. (1999) Factors involved in the prognosis of equine laminitis in the UK. *Equine Veterinary Journal* **31**, **5**, 433–442.

Denoix, J-M. and Pailloux, J-P. (1996) *Physical Therapy and Massage for the Horse*. Manson Publishing Ltd, London, UK.

Dobson, H. and Smith, R.F. (2000) What is stress, and how does it affect reproduction? *Animal Reproduction Science* **60–61**, 743–752.

Ecke, P. and Hodgson, D.R. (1996) Survey of the incidence of acute colitis in adult horses in Australia. *Australian Equine Veterinarian* **14**, **4**, 180–181.

Ecker, G.L. and Lindinger, M.I. (1995) Water and ion losses during the cross country phase of eventing. *Equine Veterinary Journal* **20**, 111–119.

Fanselow, M.S. and Baaches, M.P. (1982) Conditioned fear-induced opiate analgesia on the formalin test: evidence for two aversive motivational systems. *Learning and Motivation* **13**, 200–221.

Ford, E.J.H. (1973) Clinical aspects of ragwort poisoning in horses. *Veterinary Annual* **14**, 86–88.

Fowler, M.E. (1980) Veterinary problems during endurance trail rides. *Proceedings of the Annual Convention of the American Association of Equine Practitioners* **25**, 469–478.

Fraser, A.F. (1992) *The Behaviour of the Horse*. C.A.B. International, Oxon, UK.

Garlinghouse, S.E. and Burrill, M.J. (1999) Relationship of body condition score to completion rate during 160 km endurance rides. *Equine Veterinary Journal* **30, suppl.**, 591–595.

Gerring, E.L. and Davies, J.V. (1982) Fracture of the tibial tuberosity in a polo pony. *Equine Veterinary Journal* **14**, **2**, 158–159.

Gibson, T.E. and Paterson, D.A. (eds.) (1985) *The Detection and Relief of Pain in Animals. British Veterinary Association*. Animal Welfare Symposium, British Veterinary Association, London, UK.

Goodwin, D. (1999) The importance of ethology in understanding the behaviour of the horse. *Equine Veterinary Journal* **28, suppl.**, 15–19.

Gray, J.A. (1987) *The Psychology of Fear and Stress*, 2nd edition. Cambridge University Press, UK.

Guthrie, D.M. (1980) *Neuroethology: An Introduction*. Blackwell Scientific Publications, Oxford, UK.

Hall-Patch, P.K., Orton, R.G. and Sampson, J.H. (1977) Competitive trail and endurance riding in the UK. *Veterinary Record* **100**, **10**, 192–194.

Harman, J.C. (1995) Practical saddle fitting. *The Equine Athlete* **8**, **2**, **1**, 12–13.

Harris, P.A. (1999) Review of equine feeding and stable management practices in the UK concentrating on the last decade of the 20th century. *Equine Veterinary Journal* **28, suppl.**, 46–54.

Hertsch, B. (1992) The physical appearance in stress on the musculoskeletal system of dressage, jumping and military (three day event) horses. *Deutsche Tierarztliche Wochenschrift* **99**, **1**, 36–39.

Hinchcliff, K.W., Kohn, C.W., Geor, R., McCutcheon, L.J., Foreman, J., Andrews, F.M., Allen, A.K., White, S.L., Williamson, L.H. and Maykuth, P.L. (1995) Acid-base and serum biochemistry in horses competing at a modified 1 star three-day-event. *Equine Veterinary Journal* **20, suppl.**, 105–110.

Hobo, S., Oikawa, M.A., Kuwano, A., Yoshida, K. and Yoshihara, T. (1997) Effect of transportation on the composition of bronchoalveolar lavage fluid obtained from horses. *American Journal of Veterinary Research* **58**, **5**, 531–534.

Hodgson, D.R. and Rose, R.J. (1987) Effects of a nine month endurance training programme on muscle composition in the horse. *Veterinary Record* **121**, **12**, 271–274.

Holliday, T.A. (1972) The nervous system. In Catcott, E.J. and Smithcors, J.F. (eds.), *Equine Medicine and Surgery*, Second edition. American Veterinary Publications Inc., Wheaton, Illinois, USA.

Holmstrom, M., Fredricson, I. and Drevemo, S. (1995) Biokinematic effects of collection on the trotting gaits in the elite dressage horse. *Equine Veterinary Journal* **27**, **4**, 281–287.

Houpt, K.A. and Fregin, G.F. (1992) Equine workshop. In: *The Well-being of Agricultural Animals in*

Biomedical and Agricultural Research. Proceedings of the SCAW-sponsored conference, Agricultural Animals in Research. Scientists Center for Animal Welfare, Washington DC, USA.

Houpt, K.A. and Feldman, J. (1993) Animal Behaviour Case of the month. *Journal of the American Veterinary Medical Association* **203**, **9**, 1279–1280.

Houpt, K.A. and Lein, D. (1980) Equine Behaviour: The sexual behaviour of stallions. *Equine Practice* **2, 5**, 8–22.

Hughes, J.P., Stabenfeldt, G.H. and Evans, J.W. (1975) The oestrus cycle of the mare. *Journal of Reproduction and Fertility* **23, suppl.**, 161–166.

Hyland, J.H. (1993) Uses of gonadotrophin-releasing hormone (GnRH) and its analogues for advancing the breeding season in the mare. *Animal Reproduction Science* **33**, 195–207.

Iggo, A. (1993) Somesthetic sensory systems. In Swenson, M.J. (ed.), *Dukes Physiology of Domestic Animals*, 11th edition, Cornell University Press, Ithaca, USA.

Ito, S., Fujii, Y., Uchiyama, T. and Kaneko, M. (1992) Four cases of rhabdomyolysis in the Thoroughbred during transportation. *Bulletin of Equine Research Institute* **29**, 1–5.

Jeffcott, L.B. and Field, J.R. (1985) Epidemiological aspects of hyperlipaemia in ponies in South Eastern Australia. *Australian Veterinary Journal* **62, 4**, 140–141.

Jeffcott, L.B. (1979) Back problems in horses – a look at past, present and future progress. *Equine Veterinary Journal* **11, 3**, 129–136.

Jeffcott, L.B., Holmes, M.A. and Townsend, H.G.G. (1999) Validity of saddle pressure measurements using force-sensing array technology – preliminary studies. *Veterinary J.* **158, 2**, 113–119.

Johnson, K.G., Tyrell, J., Rowe, J.B. and Pethick, D.W. (1998) Behavioural changes in stabled horses given non-therapeutic levels of virginamycin as Founderguard. *Equine Veterinary J.* **30, 2**,139–143.

Juarbe-Diaz, S.V., Houpt, K.A. and Kusunose, R. (1998) Prevalence and characteristics of foal rejection in Arabian mares. *Equine Veterinary J.* **30, 5**, 424–428.

Kaplan, R.M. and Little, S.E. (2000) Controlling equine cyathostomes. *Compendium on Continuing Education for the Practicing Veterinarian* **22, 4**, 391–396.

Kohn, C.W., Hinchcliff, K.W., McCutcheon, L.J., Geor, R., Foreman, J., Allen, A.K., White, S.L., Maykuth, P.L. and Williamson, L.H. (1995) Physiological responses of horses competing at a modified 1 star three-day-event. *Equine Veterinary J.* **20, suppl.**, 97–104.

Lane J.G. and Mair T.S. (1987) Observations on headshaking in the horse. *Equine Veterinary J.* **19**, 331–336.

Langlois, B. and Blouin, C. (1996) Effect of a horse's month of birth on its future performances. In *Proceedings of the 47th Annual Meeting of the European Association for Animal Production*. Hungary.

Lauk, H.D. and Kreling, I. (1998) Surgical treatment of kissing spines syndrome – 50 cases. *Pferdeheilkunde* **14, 2**, 123–130.

Lawrence, L., Jackson, S., Kline, K., Moser, L., Powell, D. and Biel, M. (1992) Observations on body weight and condition of horses in a 150 mile endurance ride. *J. Equine Veterinary Science* **12, 5**, 320–324.

Leblond, A., Villard, I., Leblond, L., Sabatier, P. and Sasco, A.J. (2000) A retrospective evaluation of the causes of death of 448 insured French horses in 1995. *Veterinary Research Communications* **24, 2**, 85–102.

Leyland, A. (1985) Ragwort poisoning in horses. *Veterinary Record* **117, 18**, 479.

Lindner, A. and Schoneseiffen, N. (1996) Frequency of diseases of pleasure horses and dressage horses (in Germany). *Pferdeheilkunde* **12, 4**, 715–716.

Lucke, J.N. and Hall, G.M. (1980) Long distance exercise in the horse: Golden Horseshoe Ride 1978. *Veterinary Record* **106, 18/19/20**, 405–407.

Lyon, S., Stebbings, H.C. and Coles, G.C. (2000) Prevalence of tapeworms, bots and nematodes in abattoir horses in south-west England. *Veterinary Record* **147, 16**, 456–457.

MacLeay, J.M., Sorum, S.A., Valberg, S.J., Marsh, W.E. and Sorum, M.D. (1999) Epidemiologic analysis of factors influencing exertional rhabdomyolysis in Thoroughbreds. *American J. Veterinary Research* **60, 12**, 1562–1566.

Madigan, J.E. (1996) Evaluation and treatment of headshaking syndrome. *Proceedings of the 1st International Conference on Equine Clinical Behaviour*. Basel, Switzerland.

Madigan, J.E. and Bell, S.A. (1998) Characterisation of headshaking syndrome – 31 cases. *Equine Veterinary J.* **27, suppl.**, 28–29.

Madigan, J.E., Kortz, G., Murphy, C. and Rodger, L. (1995) Photic headshaking in the horse: 7 cases. *Eq. vet. J.* **27**, 306–311.

Mair, T.S. and Lane, J.G. (1990) Headshaking in horses. *In Practice* **9**, 183–186.

Marcella, K.L. (1990) Dressage related lameness. *Proceedings of the Annual Convention of the American Association of Equine Practitioners* **35**, 559–566.

Marcella, K.L. (1991) Polo Pony Injuries. *Proceedings of the 36th Annual Convention of the American Association of Equine Practitioners*, Lexington, Kentucky, December 2–5, **36**, pp. 647–660.

Marlin, D.J., Harris, P.A., Schroter, R.C., Roberts, C.A., Scott, C.M., Orme, C.E., Dunnett, M., Dyson, S.J., Barrelet, F., Williams, B., Marr, C.M. and Casas, I. (1995) Physiological, metabolic and biochemical responses of horses competing in the speed and endurance phase of a CCI four star three-day-event. *Equine Veterinary J.* **20, suppl.**, 37–46.

Mason, G. (1991) Stereotypies: a critical review. *Animal Behaviour* **41**, 1015–1037 .

Mayhew, I.G. (1992) How I treat headshakers. *Proceedings of the North American Veterinary Conference*, Orlando, Florida **6**, 453–454.

McDonnell, S.M. (2000) Reproductive behaviour of stallions and mares: comparison of free-running and domestic in-hand breeding. *Animal Reproduction Science* **60–61**, 211–219.

McGreevy, P. (1996) *Why does my horse . . . ?* Trafalgar Square Publishing, North Pomfret, Vermont, USA.

McLean, J.G. (1973) Equine parasitic myoglobinuria ('azoturia'): a review. *Australian Veterinary J.* **49, 1**, 41–43.

Melzack, R. and Wall, P.D. (1988) *The Challenge of Pain*, 2nd edition. Penguin Books, London, UK.

Mills, D.S. and Nankervis, K.J. (1999) *Equine Behaviour: Principles and Practice*. Blackwell Science, Oxford, UK.

Mills, D.S., Cook, S., Jones, B. and Taylor, K. (2001a) A demographic analysis of 254 cases of equine headshaking with particular reference to their differential symptomatology. *Veterinary Record* (In press)

Mills, D.S., Cook, S., Jones, B. and Taylor K. (2001b) Reported response to treatment amongst 245 cases of equine head shaking. *Veterinary Record* (In press)

Morales, J.L., Manchado, M., Cano, M.R., Miro, F. and Galisteo, A.M. (1998a) Temporal and linear kinematics in elite and riding horses at the trot. *J. Equine Veterinary Science* **18, 12**, 835–839.

Morales, J.L., Manchado, M., Vivo, J., Galisteo, A.M., Aguera, E. and Miro, F. (1998b) Angular kinematic patterns in elite and riding horses at trot. *Equine Veterinary J.* **30, 6**, 528–533.

Moritsu, Y., Terai, A., Kimura, M. and Ichikawa, S. (1998) The effect of the month of birth on racing performance in Thoroughbred racehorses. *J. Rakuno Gakuen University Natural Science* **23, 1**, 1–4.

Newton, S.A., Knottenbelt, D.C. and Eldridge, P.R. (2000) Headshaking in horses: possible aetiopathogenesis suggested by the results of diagnostic tests and several treatment regimes used in 20 case. *Equine Veterinary J.* **32**, 208–216.

Personett, L.A., McAllister, E.S. and Mansmann, R.A. (1983) Proximal suspensory desmitis. *Modern Veterinary Practice* **64, 7**, 541–545.

Raidal, S.L., Love, D.N. and Bailey, G.D. (1995) Inflammation and increased numbers of bacteria in the lower respiratory tract of horses within 6 to 12 hours of confinement with the head elevated. *Australian Veterinary J.* **72, 2**, 45–50.

Richardson, R.C. (1994) *The Horse's Foot and Related Problems: A Practical Guide for Owners, Farriers and Vets*. Greatcombe Clinic, Holne, Devon, UK.

Rose, R.J., Ilkiw, J.E., Sampson, D. and Backhouse, J.W. (1980) Changes in blood gas, acid-base and metabolic parameters in horses during three day event competition. *Research in Veterinary Science* **28, 3**, 393–395.

Rose, R.J., Purdue, R.A. and Hensley, W. (1977) Plasma biochemistry alterations in horses during an endurance ride. *Equine Veterinary J.* **9, 3**, 122–126.

Ross, W.A., Kaneene, J.B. and Gardiner, J.C. (1998) Survival analysis and risk factors associated

44 R.A. CASEY

with the occurrence of lamenss in a Michigan horse population. *American J. Veterinary Research* **59, 1**, 23–29.

Schatzmann, U. and Klemm, P. (1998) Health problems in polo ponies. A survey. *Pferdeheilkunde* **14, 6**, 478–484.

Schmidt, B. and Schmidt, K.H. (1980) Effect of road transport, lunging, competition and time of day on activities of serum enzymes aspartate aminotransferase, creatinine kinase, lactate dehydrogenase, alkaline phosphate and serum bilirubin in warm blooded horses. *Berliner und Munchener Tierarztliche Wochenschrift* **93, 13**, 244–246.

Schott, H.C., McGlade, K.S., Hines, M.T. and Petersen, A. (1996) Bodyeight, fluid and electrolyte, and hormonal changes in horses that successfully completed a 5 day, 424 kilometre endurance competition. *Pferdeheilkunde* **12, 4**, 438–442.

Seligman, M.E.P. (1970) On the generality of the laws of learning. *Psychological Review* **77**, 406–418.

Slade, L.M. and Branscomb, J. (1989) Conformation traits of jumping horses. Proceedings, Western Section, *American Society of Animal Science and Western Branch Canadian Society of Animal Science* **40**, 35–38.

Smith, B.L., Jones, J.H., Carlson, G.P. and Pascoe, J.R. (1994) Body position and direction preferences in horses during road transport. *Equine Veterinary J.* **26, 5**, 374–377.

Snow, D.H., Baxter, P. and Rose, R.J. (1981) Muscle fibre composition and glycogen depletion in horses competing in an endurance ride. *Veterinary Record* **108, 17**, 374–378.

Stull, C.L. and Rodiek, A.V. (2000) Physiological responses of horses to 24 hours of transportation using a commercial van during summer conditions. *J. Animal Science* **78, 6**, 1458–1466.

Sweeney, C.R., Divers T.J. and Benson, C.E. (1985) Anaerobic-bacteria in 21 horses with pleuropneumonia. *Journal of the American Veterinary Medical Association* **187, 7**, 721–724 .

Swenson, M.J. (ed.) (1993), *Dukes Physiology of Domestic Animals*, 11th edition. Cornell University Press, Ithaca, USA.

Tong, J.M.J. (1994) Acute rupture of the Corpora nigra caused by a traumatic incident to the left eye of a polo pony. *Equine Veterinary Education* **6, 3**, 118–121.

Trevillian, C., Holt, J. and Yovich, J. (1997) Evaluation of cardiac recovery index and clinicopathological parameters in endurance horses. *Australian Equine Veterinarian* **15, 2**, 83–88.

Ulmer, R.S. (1985) Hormonal control of the reproductive cycle in the mare. *Southwestern Veterinarian* **36, 2**, 101–105.

Vatistas, N.J., Snyder, J.R., Carlson, G., Johnson, B., Arthur, R.M., Thurmond, M., Zhou, H. and Lloyd, K.L.K. (1999) Cross-sectional study of gastric ulcers of the squamous mucosa in Thoroughbred racehorses. *Equine Veterinary Journal* **29, suppl.**, 34–39.

Voynick, B.T. and Sweeney, C.R. (1986) Exercise induced pulmonary haemorrhage in polo and racing horses. *Journal of the American Veterinary Medical Association* **188, 3**, 301–302.

White, S.L., Williamson, L.H., Maykuth, P.L., Cole, S.P. and Andrews, F.M. (1995) Heart rate response and plasma lactate concentrations of horses competing in the speed and endurance phase of three-day combined training events. *Equine Veterinary J.* **20, suppl.**, 52–56.

Whitwell, K.E., Harris, P. and Farrington, P.G. (1988) Atypical myoglobinuria: an acute myopathy in grazing horses. *Equine Veterinary J.* **20, 5**, 357–363.

Willms, F., Rohe, R. and Kalm, E. (1999) Genetic analysis of different traits in horse breeding by considering radiographic findings. 2: Genetic relations between conformation traits, performance traits, and bone diseases. *Zuchtungskunde* **71, 5**, 346–358.

Woods, G.L. and Steiner, J.V. (1986) Embryo transfers from mares in athletic competition. *Cornell Veterinarian* **76, 2**, 149–155.

Chapter 3

NUTRITION AND WELFARE

N. DAVIDSON and P. HARRIS
Equine Studies Group, WALTHAM Centre for Pet Nutrition, Freeby Lane, Waltham-on-the-wolds, Leicestershire LE14 4RT, UK

Abstract. The horse is a social species living in herds and spending the majority of its time roaming and foraging in a diverse and seasonally-varying environment. As a non-ruminant herbivore it is well suited to a high fibre, low starch diet. Domestication has resulted in a number of benefits to the horse, reflected in its continued prevalence and apparently increased life expectancy, but it has not been without its price. Especially in developed countries, horses kept for leisure purposes (which includes all competition and racing horses) are often confined, possibly away from conspecifics, within a stable for a large proportion of the day. Due to increased energy requirements many horses now receive one to two large meals a day, consisting of feedstuffs with a low water content and often a radically different nutritional profile from the diet that they would be able or would choose to select in the wild. These modern practices have benefits but also potential disadvantages to the horse both nutritionally and behaviourally which may have an impact on welfare. This chapter highlights areas where dietary imbalances or inappropriate feeding practices may potentially have an adverse effect on welfare and gives suggestions on how these may be ameliorated.

1. Introduction

Since the horse was first domesticated, it has been kept for a variety of purposes including for meat, as a means of transport and for leisure pursuits. This variability still exists today, depending especially on the type of horse and where it is to be found in the world. The way horses are kept and managed also varies considerably both within and between countries and reflects the purpose for which they are kept, where they are kept, what is available, the time of year, their breed/age, as well as the owner's financial situation (Harris 2000). In order to assess whether current feeding practices potentially have a positive or negative effect on welfare, it is important to determine what we are comparing them with. For example our improved knowledge of nutritional requirements, along with our increased knowledge of veterinary medicine, contributes to the fact that many horses today, overall, may be healthier for longer than either their wild ancestors or horses at the beginning of the last century. However, many of our modern management practices (*e.g.* feeding large amounts of cereals in a small number of meals) have a potentially negative impact, and may be compromising today's horses in ways not associated with their ancestors.

It is always important to remember that compared with the horse in the wild, we are now responsible for both our horse's diet and the effect that it may have on health and behaviour. This chapter will discuss some of the major areas of nutrition, which have the potential to affect welfare negatively.

45

N. Waran (ed.), The Welfare of Horses, 45–76.

2. Natural Diet and Feeding Behaviour

In order to understand the influence that modern feeding and management practices may have, it is helpful to consider the natural diet and feeding behaviour of the horse. The modern horse evolved essentially as a plains dweller, ranging up to 80km per day and exploring and roaming over wide open spaces. Horses do not ruminate and are often referred to as 'trickle feeders' (Harris 1999a) as they naturally ingest relatively small amounts of feed at each bite, take a few bites and then chew. They will then move and start the cycle of biting and mastication again. They tend not to eat a discrete meal from one plant or to concentrate solely upon one species of plant, even if it is a preferred plant species (Mariner 1980). The diet of free ranging horses has been shown to be very varied, both on a daily basis and from season to season (Hansen 1976; Ralston 1984; Gill 1988) and because of this, the feeding behaviour of free ranging horses has been thought to be more complex than its stabled counterpart (Carson & Wood-Gush 1983). Although horses prefer to graze, New-Forest ponies take up to 20% of their diet as browse species even at the peak time for grass (Putman et al., 1987; Gill 1988). Similar findings were seen with some American feral horse populations (Hansen 1976).

Horses on managed pastures still show some species preferences (Archer 1971, 1973), although the preferences found in one study are not necessarily confirmed in others and this will reflect the species available, time of year and the methodologies employed (Archer 1978; Avery 1996; Nash 2001). The nature of the pasture affects the amount that is ingested with every bite, with more herbage being ingested with dense high leaf to stem ratio. In order to compensate for a reduced quality of pasture horses, have been shown to increase both their grazing time and their bite frequency (Rolgalski 1970; Nash & Thompson 2001). Whether free-living or domesticated, horses tend to spend around 16–18 hours out of every 24 hours foraging, depending on the type of grazing available (Avery 1996; Nash & Thompson 2001). They rarely fast voluntarily for more than 4 hours at a time. Both stabled and free ranging horses spend a significant amount of their time foraging at dawn, dusk and night (Tyler 1972; Ruckebusch et al., 1976).

3. The Digestive System

The horse's digestive system is well suited to its almost continual intake of forage ('trickle feeding'). An overview of the digestive process in the horse, and how diet may influence it, is given below (for further details see Lewis 1995; Frape 1998; Harris 1999a).

The horse selects its preferred food items using its prehensile lips, tongue and teeth. The jaw movements involved in chewing are complex and incorporate both lateral and vertical components. It has been estimated, that (in a 500 kg horse) the jaw makes over 50,000 complete movements a day when grazing. The jaw movements at grass are relatively wide and long but when eating hay, and in particular cereals or pelleted feeds, the movement is confined. The nature of the feeds fed will therefore dramatically influence the chewing rate and speed of ingestion. The horse does not salivate at the sight of food, only producing saliva while it is actually chewing. The more a food item is required to be chewed, the more saliva that is produced and mixed with the feed, which in turn makes the bolus more moist (thus lubricating its passage to the stomach).

The stomach volume of the adult horse (500 kg bodyweight) is relatively small, around 9–15 litres. The stomach is relatively inelastic and has a finite capacity. The musculature is such that vomiting or gastric reflux is extremely rare and usually reflects a significant abnormality. The rate of gastric emptying is largely dependent on the square root of the volume, so that the larger the meal, the more rapid the rate of gastric emptying and the faster the food passes through the small intestine.

The stomach is divided into two sections, which have both anatomical and physiological differences. Food enters from the oesophagus into the cranial non-glandular section where bacterial fermentation of the ingested feed starts. This mainly involves lactobacteria, which convert any available simple sugars or starches to lactic acid. This microbial activity and degradation is stopped when the gastric contents pass to the fundic gland region and mix with the acid stomach juice. When large concentrate meals are fed, the swallowed bolus has a high dry matter (DM) content and the stomach contents therefore also have a higher DM content than if the horse was grazing. This results in slower and /or reduced mixing of the feed with the gastric juices and may allow the survival of, in particular, gas-producing bacteria (Meyer *et al.*, 1975; Harris 1999b). This increases the risk of disturbances due to fermentation and increased gas production. Whether this results in a clinical or sub-clinical problem will depend on the actual amount eaten and the rate of intake, the amount of available sugars and starches and the microbial population.

The basic digestive processes of the small intestine (enzymatic degradation of proteins, fats, starches and sugars) are similar to those of other monogastric animals *but* the activity of some of the enzymes in the chyme, in particular amylase, is less than in other monogastric animals. The horse, therefore, has a limited capacity to digest starch but a comparatively high capacity to digest sugars in the small intestine (Cuddeford 1999a).

Large starch or sugar based meals may overwhelm the digestive capacity of the stomach and small intestine, leading to the rapid fermentation of the grain carbohydrate in the hindgut and a decrease in pH. A significantly decreased caecal pH may initiate a chain of events, which include a change in the microbacterial flora (excessive growth of those bacteria that can live under such conditions), lysis of those bacteria which cannot live at such low pH, allowing the release of endotoxins and other molecules, as well as damage to the mucosa of the caecum and colon. This in turn may allow the absorption of endotoxins and various other materials with potential clinical consequences, including colic, diarrhoea and laminitis (Frape 1988; Clarke *et al.*, 1990; Lewis 1995; Harris 1999a). Hay or roughage based diets do not result in such decreases in caecal pH (Willard *et al.*, 1977) but may not provide sufficient net energy for many horses' needs (Harris 1997a).

There are large fluid movements associated with food digestion. Chewing and consequent salivation result in fluid being drawn from the circulating blood. In horses that chew slowly this fluid can easily be compensated for but in the avaricious feeders the changes in the circulating plasma can be fairly marked and equivalent to those found with dehydration. Restoration may take up to a few hours. This is one of the reasons why encouraging a moderate intake rate is desirable and also why exercising horses soon after a large meal may not be ideal. More than 100 litres of fluid may be secreted into the pre-caecal section of the gastrointestinal tract per day in a typical 500 kg horse. In addition, in discrete meal-fed horses large volumes of fluid are initially secreted into the hind gut as the chyme from the small intestine reaches there, followed by a period of resorption. In horses fed in a more natural way, *i.e.* as trickle feeders, the fluid shifts are much less marked in the hindgut and there are also far less marked fluctuations in the hind gut microflora. The horse relies heavily on sweating for thermoregulation in high temperatures and during exercise and unwanted nitrogen has to be removed as urea in the urine, which also requires water. These together with the large fluid secretions into the gastrointestinal tract help to explain why a constant supply of clean water is essential for the horse.

4. Current Management Practices

Few horses in the developed world are kept today for agricultural purposes or for transportation. Most are kept fundamentally for leisure purposes and two major factors influence their management. Firstly whether they are kept inside in stables/barns or outside on pastures, paddocks or dry-lots and secondly their energy requirements (see Harris 2000).

In a Scottish survey (Mellor *et al.*, 1997), only 10% of horses were kept permanently out at pasture (~10,000 horses), 29% were stabled most of the time and a further 2% were permanently stabled. In a UK based survey (Anon-BETA 1999), it was estimated that approximately 145,000 horses were kept at grass, all year round. In the riding schools and livery yards 23% of horses that were owned by these establishments were stabled all year and 16% were kept in the open all year, as compared with 29 and 11% respectively of those horses owned by others but kept at livery at these establishments. The most common practice was to keep horses stabled in the winter only.

Most horse pastures in the developed world are managed. Well managed and fenced pastures that are not overgrazed and that contain suitable plant species may be able to provide most if not all of the nutrient needs of the horses which graze them (Avery 1996; Nash *et al.*, 2001). If combined with social groups of conspecifics these can also satisfy many of the behavioural needs of horses as well. Unfortunately many pastures do not reach this ideal and for climatic and geographical reasons will not support the nutritional requirements of all horses throughout the whole year, especially horses that are growing or exercising, even if the pastures are well managed (Avery 1996; Kronfeld *et al.*, 1996, 1998; Nash 2001). Overgrazing and inadequate endoparasitic control can also be detrimental to the health of horses. In the wild, horses roam over large areas but when kept on relatively small pastures and allowed to graze freely, horses will quickly establish patches within the pasture. This will include areas that they use for grazing and large patches that they use for excretion (this applies particularly to mares, fillies and geldings and is different to the dung piling that stallions participate in as marking behaviour). It has been estimated that, on a pasture previously ungrazed by horses, around 25–75% of the available sward may be grazed, depending on the density of stocking, the remainder being used for excretion (Archer 1978). If left unmanaged this habit of pattern grazing by horses can lead to the grazed areas being impoverished, bare and potentially infested with unwanted or even poisonous plants such as ragwort (*Senico jacobaea*). The ungrazed areas become coarse and unpalatable even to cattle and overgrown with weeds. Proper pasture management and in particular faeces removal is vital when horses are kept on relatively small pastures (Avery 1996). Optimum stocking densities will depend on the quality of pasture, but as a general rule of thumb, the recommendation is that at least two acres are provided for the first horse, and an extra acre for every additional horse. Time budgets may be affected according to the nature of the pasture – fillies are much more active when grazed on improved pastures than when on unimproved pastures (Nash & Thompson 2001) as potentially less time is required to be spent grazing.

Pastures established for other herbivores such as cattle, may be equally adverse (although combined cattle and horse grazing is recommended to be beneficial (see Avery 1996) and inadequately fenced and maintained pastures may actually prove hazardous. Stabling horses has the advantage that it provides protection against environmental extremes and ectoparasites, provision for individual feeding or care as required, as well as reducing the risk of accidental damage by other horses. However, by confining horses to a stable we not only restrict their movement and social contact with other horses but also affect their perceived ability to remove themselves from danger.

Domestication, and our increasing demand for horses to perform repeatedly has resulted in energy requirements that, for some horses, are above those able to be provided by their 'natural' diet of fresh forage. Cereals provide more net energy than hay, which in turn provides more than twice the net energy of straw (Martin Rosset *et al.*, 1994; Harris 1997a) which has resulted in their inclusion in the diets of many horses (Harris 1997b, 2000). Confinement within a stable is therefore often coupled with the feeding of highly concentrated cereal-based meals a few times a day with limited forage (Harris 1997b). However, the feeding of discrete concentrate based meals is not confined to the stabled horse as many horses out at pasture are still fed concentrates, especially the gestating and lactating mare and growing animals.

5. Welfare Implications

A balanced and appropriate diet is crucial for the optimal health and welfare of any horse. Welfare issues often arise when overt deficiencies and/or toxicities, such as with selenium or vitamin D, occur sometimes through neglect, ignorance or a formulation error (N.R.C. 1989; Lewis 1995; Harkins *et al.*, 1997; Frape 1998). Similarly, the ingestion of poisonous plants (such as ragwort (*Senecio* spp.) and yew (*Taxus* spp.), or feed contaminated with toxic components (such as monensin, which can be included in cattle feed as a growth promoter), or certain fungi/bacteria or their associated toxins (such as the ergot alkaloids produced by an endophyte fungus (*Acremonium coenophialum*) found on fescue, or botulism associated with big bale silage) or insects (such as blister beetles) can affect the horse (Lewis 1995; Harkins *et al.*, 1997; Frape 1998; Knight & Walter 2001). These are outside the scope of this chapter but should always be considered in the wider context of nutrition and health.

As outlined above most horses in the developed world are kept for leisure – some for the individual owner's pleasure or competition purposes, others are part of a commercial enterprise, for example a riding school or a thoroughbred stud. The pressures on each of these different enterprises varies considerably – for the riding school horses must be capable of undertaking many hours of relatively repetitive work with different riders on board – for the thoroughbred stud it may be to raise youngstock for the sales or for racing. The latter can produce its own problems, as weanlings and yearlings are well muscled tend to attract the highest prices at sales, which often encourages the production of 'fat' animals. This may result in health problems in the young animal but may also have carryover effects on its health as an adult. The term 'production disease' is applied to man-made diseases of live-stock derived from breeding, feeding and management for high production *e.g.* milk fever, mastitis and laminitis. Certain conditions in the horse, especially Developmental Orthopaedic Diseases (DOD, see below) are considered to be pro-duction diseases (Kronfeld 1997). Other relatively new conditions, which have a genetic component, have become prevalent because of the demand for the offspring of the originating animal. The most obvious example of this is the condition of hyperkalaemic periodic paralysis (HYPP, an incompletely dominant single auto-somal gene disorder) which appears to have originated as a point mutation in a stallion called 'Impressive' which was in demand because of the type and success of the offspring that he produced (Naylor *et al.*, 1993). Nutrition (providing a low potassium dietary intake) can be used in this particular condition to help reduce the incidence and severity of the episodes (Topliff 1997). Similarly other medical conditions, not necessarily with a primary nutritional aetiology, may be helped by appropriate dietary management. This is the case for Equine Rhabdomyolysis Syndrome (often referred to as 'tying-up') (Harris 1999b).

Managemental practices vary considerably throughout the world and present their own unique welfare issues, for example, Equine Motor Neurone Disease in horses kept on dry-lots with no access to green forage (which is believed to be associated with inadequate Vitamin E intake (Valentine 1997)) or sand colic (see below) for those kept on sandy dry lots or sandy pastures.

Some feeding practices that appear on the surface to be beneficial may have disadvantages. In 1994, Brown and Powell Smith reported that many people in the UK prefer to feed hay on the stable floor which may be considered to be a more natural way of feeding (although horses naturally feed at different heights depending on the grazing and browse available). This was endorsed by Townson's survey of a number of Irish racing stables (Townson *et al.*, 1995) which noted that 57% of the racing stables fed hay from the floor. The disadvantages of hay racks or nets high from the ground have been suggested as including an increased risk of particles getting into the eyes and nose, as well as, feet getting caught in the bars/hay net. As this is not the natural feeding position it has been suggested to adversely affect drainage within the respiratory system, to increase the risk of developing caudal and cranial dental hooks and perhaps to adversely affect muscle and nerve function (Hintz 1997). Hay racks at chest height are thought to have the potential to increase the risk of injuries, decrease the space available within the stable and are costly. However, despite hay feeding on the floor being a more natural feeding posture, it increases wastage by increasing the risk of contamination with faeces and urine, and there is an increased risk of parasite egg ingestion.

Finally it should be remembered that some horses are neither kept in discrete paddocks or horse stables but are kept wherever there is space, even if that is on waste tips or in garages. These animals, such as the urban ponies in Dublin, present unique welfare issues that are outside the scope of this chapter.

6. Optimising Diet and Managemental Practices

The closer a feeding system programme can get to the natural system of feeding, the easier it is to maintain gastrointestinal tract homeostasis. But we have to accept that today it is often not possible to keep all horses in the natural way and still maintain their current role in society. There are, however, a number of relatively simple feeding and management practices that may help reduce the problems associated with current methods of managing horses.

6.1. DENTAL PROBLEMS

Horses adult teeth grow continuously throughout their life, gradually being ground down through chewing fibrous, silica-containing forage as well as the kernels of hard grains. The circular jaw movements undertaken when chewing grasses are wider and produce more even wear across the chewing surface, than when preserved forages are being fed. The movements are even more restricted with grains and manufactured feeds. This smaller movement together with the normal anatomy of the equine head (where the upper dental arcade is wider than the lower), results in sharp edges and hooks developing more quickly on the upper lateral, and lower medial teeth surfaces. This can cause 'quidding' (dropping feed while chewing), lacerations within the mouth, pain and may eventually lead to loss in body condition (Dixon 2000). Most other dental problems, not relating to hooks or general wear and tear, occur early on in life. Some are congenital such as Parrot mouth (where the lower jaw is considerably shorter than the upper jaw) or Sow mouth (the opposite to Parrot mouth and where the upper jaw is shorter than the lower jaw).

Alternatively they may result from an accident or an unwanted stereotypic behaviour such as cribbing or wood chewing. Other oral irregularities may occur which are initially easily correctable but over a period of time the relentless grinding action against the irregularity results in a reshaping of the dental arcade into a much more severe abnormality. In the older horse periodontal disease and decay are relatively common. Indicators of teeth problems include quidding, being reluctant or slow to eat the ration, and 'bit' resistance (Pilliner 1996; Meszoly 2001).

6.1.1. *Reducing dental problems*

A recent study has suggested that dental correction in horses without severe dental abnormalities may not have a major effect on digestive efficiency (Ralston *et al.*, 2001), but clinically it has been reported to be of benefit with respect to weight gain and cessation of quidding in those with more severe abnormalities (Dixon 2000). Horses teeth should be regularly checked by a veterinarian, perhaps as frequently as every 6 months, depending on the individual horse and the anatomy of its jaws, the evenness of its dental arcade and its diet. Typically, unless there are known problems, horses between 5 to 15 years of age may only require an annual oral examination whereas younger and older animals, as well as those with problems, will need to be evaluated on a 6 monthly or more frequent basis. In between the regular inspections it is advisable to regularly observe the individual horse's eating pattern so that changes in feeding behaviour can be noted, and it is helpful to be alert to bitting problems. In particular, it can be useful to look out for the presence of any unusual swelling on the upper or lower jaw, excessive salivation, quidding, unusual tilting of the head whilst chewing, unpleasant odour from the mouth or nostril or unusually high amounts of long fibres and grains in the faeces. Catching dental abnormalities in the early stages is preferable to dealing with the long term consequences.

6.2. CHOKE

The typical DM content of a swallowed grass bolus is around 11%, that of a hay bolus is around 20%, whereas that of a cereal-based feed bolus is more like 30–40%. The occurrence of oesophageal obstructions (choke) depends not only on the swelling capacity of the feedstuff but also the speed of feed intake and the size, nature and DM content of the boluses swallowed. Choke has been said to be most commonly due to impaction with pieces of sugar beet, apple or carrots or pelleted/heavily processed feed (Feige *et al.*, 2000). The risk is thought to be increased in horses, which are unable to chew adequately due to old age or poor dentition, or those, which bolt their feed. These factors can be assisted by appropriate management practices.

6.2.1. *Reducing the risk of choke*

The grinding of whole grains is necessary for their optimal digestion in the small intestine. The intensity of the grinding of the roughage may be important for the passage of digesta through the ileocaecal-colic junction into the large intestine.

Therefore, regular dental care is essential to maintain optimal mastication, and consequently optimal digestion. Steps should be taken to prevent horses from bolting their food, which include (Hintz 1994):

- Adding short chopped fibre or chaff (> 2 cm in length);
- Feeding smaller meals more frequently;
- Spreading the feed thinly over a large surface;
- Feeding nervous horses first;
- Adding large, smooth, stones to the feeding trough;
- Splitting the meal into many small compartments within the feeding trough or manger area;
- Using coarse mixes or extruded feeds rather than pelleted feeds.

Lawn clippings or other fine grasses, should be avoided as well as indigestible fibres including corn cobs, twigs, and very mature stemmy forage especially for those horses that are prone to choke or have reduced ability to chew adequately. Sugar beet should be well soaked and each animal allowed to eat the meal at its own pace and pattern (Marie 1999).

6.3. GASTRO-INTESTINAL DISTURBANCES

The rapid digestion of a concentrated meal has been shown to cause distinct physiologic disturbances when compared with grazing or steady state feeding conditions (Williard *et al.*, 1977; Clarke *et al.*, 1990; Pagan *et al.*, 1999). These include:

- Fluctuations of plasma glucose and metabolic hormones;
- Postfeeding increases of plasma proteins and osmolarity;
- Activation of the rennin angiotensin -aldosterone system;
- Periods of intense colonic fermentation with induction of transmural fluid secretion and reduced colonic pH.

Such episodic processes have been suggested to contribute to the incidence of digestive disorders in the stabled horse (Clarke *et al.*, 1990; Ralston 1992). Though defined as abdominal pain, colic in horses has increasingly been taken to represent the large group of intestinal diseases which cause abdominal pain, and at least 6 different types of colic have been recognised (including impaction, spasmodic and sand). In some of these conditions diet is not associated with the patho-physiology but it is frequently incriminated as the major causative factor. The exact relationship between colic and diet is difficult to determine because of the variety of feeds and feeding practices used throughout the world, as well as differences in the study populations. In addition, it is difficult to separate the effects of diet and feeding schedule from other management practices depend on the horse's breed and use. However, in one prospective study in Virginia in the USA, colic incidence was low in horses on pasture who were receiving no grain. The incidence of colic increased as the amount of concentrates fed increased, with those receiving more than 5 kg of concentrates per day being over six times more likely to develop colic than horses receiving no concentrates (Tinker *et al.*, 1996; Tinker *et al.*, 1997).

The colic risk became significant in this study when more than 2.5 kg of concentrates were fed per day. The type of concentrate fed appeared significant, with problems being particularly associated with the processed feed and pellets. Dividing large concentrate meals into several meals a day did not reduce the colic risk, in fact feeding twice daily increased the risk of colic relative to feeding once daily or feeding 3 or more times (but this may be related to the size of each respective meal). In another prospective, case control study in Texas, neither the amount or type of concentrate fed was associated with the colic risk although the researchers did conclude that horses at pasture may have a decreased risk of colic (Cohen *et al.*, 1999). Here changes in diet and in particular the type of hay fed (including hay from a different source or cutting of the same type of hay) were the key risk factors (Cohen *et al.*, 1999). In this study feeding hay other than coastal Bermuda or alfalfa significantly increased the colic risk but this may have reflected hay quality and digestibility rather than type of hay *per se*. Changing to a poorer quality, less digestible, hay or feeding wheat straw or cornstalks may predispose horses to large colon impaction regardless of the type of hay being fed (Hintz 1994; Cohen *et al.*, 1999; King 1999).

A horse that is fed excessive cereal starch or a meal that is too large or experiences a sudden change in diet (of either volume or composition), may suffer from hindgut acidosis. This can also be caused by excessive intake of rich grass that is high in water-soluble carbohydrates (*i.e.* simple sugars as well as the more complex storage carbohydrate: fructans) usually found in larger quantities during the spring and autumn. In a practitioner- based colic study in the UK, a recent change in management was associated with at least 43% of the cases of spasmodic or mild undiagnosed colic. The most common management change was turnout onto lush pasture in the spring (Proudman 1991). The ingestion of high concentrate and low forage diets has also been implicated in the development of gastric ulcers, which in turn may result in signs of colic (Murray *et al.*, 1996).

Miniature horses are particularly susceptible to small colon impaction and it has been suggested that hay quality suitable for other horses may not be sufficient for miniatures. This breed commonly has dental problems, which might be a contributory factor (King 1999). Alfalfa hay may increase the risk of enterolith formation and subsequent colic (King 1999).

Sand colic, the involuntary or voluntary ingestion of sand or small grains of soil, can be seen in particular in horses fed and/or kept on sandy pastures. The sand accumulates in particular at the pelvic and sternal flexures of the large intestine and can cause necrosis of the lining and reduced gut motility (Ruohoniemi *et al.*, 2001). Clinical signs vary but diarrhoea, colic and depression are most commonly seen (Ruohoniemi *et al.*, 2001). Poor nutrition has been implicated, but not confirmed, with the suggestion that a deficient diet may cause horses to seek nutrients from inappropriate sources. Foraging on poor pastures may cause horses to accidentally ingest sand. Studies by Lieb and Weise (1999) indicated that horses may ingest more sand when grain is fed on the ground than when grass hay is fed this way. Others have suggested that more is ingested when alfalfa hay is fed on the ground than grass hay, perhaps due to the alfalfa having a smaller leaf size and being more palatable. Weise and Lieb (2001) however, found no effect of low protein, low energy or combined low protein and low energy diets on sand intake. There was an individual effect, with some horses consuming a mean of 1361 g of sand out of the 2000 g offered. It may be that voracious voluntary sand ingestors may have learned this behaviour from having been kept on poor pastures, but there is no evidence for this. Poor gut motility associated with a high endoparasitic burden may further exacerbate the situation, reducing the transit of sand even more.

In the Texas study, Cohen et al. (1999) concluded that whilst a regular pro-gramme for administration of anthelminthics might reduce the overall frequency of colic (see also Uhlinger 1990) recent administration of anthelminthics might predispose some horses to colic. Tapeworm infection has been associated with ileal impaction and spasmodic colic in the UK (Proudman et al., 1998). However, other studies do not show such a clear link between parasite control and colic, again probably due to the study groups used, the protocols and the multiple aetiologies of colic.

6.3.1. Reducing the risk of gastrointestinal disturbances

It is important to remember that when trying to prevent colic many factors need to be considered, including, but not exclusively, the diet. These include adequate provision of water, maintenance of a regular feeding and exercise schedule, good pasture management, appropriate bedding material and endoparasite control etc.

6.3.1.1. Forage provision.

Forage (fresh or preserved) should be the foundation of any horse's diet, even those in hard work. Many horses and ponies will not require any other food. Hays with higher energy levels and greater digestibilities should be considered, especially for those animals in competitive work. Those horses in little or no work or those with an especially efficient metabolism (often called 'good doers') may benefit from being fed lower energy-containing roughages but care must be taken that this does not increase the risk of impactions. For the majority of horses, even those in work, at least 50% of their diet on a dry matter basis should be suitable forage (around 1 kg DM/100 kg BW). Even fit, very intensively working horses should be fed at least 35% and preferably 40% of their dry matter (DM) intake as forage. It may be valuable to offer some hay to horses on apparently good pastures, as the fibre content of lush pastures may not be adequate to meet their fibre needs.

Forage type should not be changed rapidly and poor quality forage should be avoided. In addition, items such as lawn clippings, large amounts of rapidly fer-mentable feeds such as apples, or feeds designed for other types of animals should not be given to horses.

Grain should only be fed when the horse's energy needs cannot be met by forage alone (sometimes small amounts of grain or other concentrates may also be needed to carry supplemental protein or minerals to balance the ration).

6.3.1.2. Alternative fibre sources.

There has been an increasing interest in the use of alternative energy sources for horses, especially alternative fibre sources which do not cause marked disturbances in the hindgut and provide more energy than typical forages. Sugar beet pulp (SBP) and soya hulls are two such fibre sources. The fibre (non-starch polysaccharide, NSP) in beet pulp is highly digestible over the total tract, with a significant proportion being degraded (approximately 16.5% of unmolassed SBP NSP) in the small intestine during transit to the hind gut (Moore-Colyer et al., 1997; Hyslop 1998). Digestibility studies suggest that not only is SBP well fermented but that degradation occurs to a large extent within the time period that such a feedstuff would remain within the gut (Stefansdottir et al., 1996; Hyslop et al., 1998). Recent studies have shown that sugar beet pulp addition to the diet may increase the nutrient value of concurrently fed hay, especially if this has a low protein content (Moore-Colyer & Longland 2001) and also may have an effect on alfalfa based diets (Hastie & Longland 2001).

6.3.1.3. *Oil.* There is an increasing use of supplementary vegetable oils for horses (Harris 1997a, b) as they provide proportionally more net energy than cereals, yet contain no starch or sugar and may provide other advantages. Although corn oil may be one of the most palatable oil, horses will vary in their preferences and providing the vegetable oil is fresh, not rancid, of a good quality, palatable and digestible to that individual, it may be acceptable. The optimal desired fatty acid composition is not yet known.

Supplemental oil should be introduced slowly since it might result in GIT disturbances. The amount of oil that should be added is still unclear. Horses have been shown to be able to digest and utilise up to 20% or more of the diet as oil. A relationship between dietary fat and muscle glycogen concentration indicated a peak glycogen at 12% oil by weight. A variety of trials have confirmed the value of incorporating this level of oil in a balanced feed (Kronfeld & Harris 1997a). However, adding oil to existing feed has the potential to create multiple imbalances, it is therefore prudent to use less than the 12% suggested above. Levels of 5–8% in the total diet are common in the competition horse. Pagan (1999) recommends 100 g/100 kg BW/day, and the majority of animals (500 kg BW) can be supplemented up to 400 ml/day (~370 g) in divided doses without any problems – provided that the oil has been introduced gradually, is required, is not rancid and the vitamin E levels are considered (Harris 1999b). In order to obtain metabolic benefits from the feeding of oil, in addition to those associated with its high energy density and lack of starch content, the oil needs to be fed for several months (Harris 1997a, Kronfeld & Harris 1997). It is recommended that additional vitamin E be fed in combination with supplemental oil, an additional 100 iu vitamin E/100 ml supplemental oil is suggested (Harris 1999b).

6.3.1.4. *Cereal.* The pre-caecal and even pre-ileal digestibility, the amount being fed, the feedstuff under consideration and the extent and nature of the processing to which it has been subjected will determine glucose and volatile fatty acid production (Kienzle et al., 1992; Kienzle 1994; DeFombelle et al., 2001). Although most people recommend that the amount of starch in each meal should be limited, there are relatively few details of what may be considered to be too large a meal

(Hintz 1994). Feeding, in a meal, around 400 g starch per 100 kg bodyweight has been suggested to saturate the small intestine's digestive capacity for starch (Potter *et al.*, 1992). Feeding less than 300 g starch per 100 kg BW was recommended by another author (Kienzle 1994) and recently it has been suggested that even at this level there may be concerns, depending on the nature of the feed (De Fombelle *et al.*, 2001), since at 300 g starch per 100 kg BW all the oat starch was digested in the small intestine, but 20% of the barley starch and 34% of the corn starch escaped the pre-caecal digestion and reached the large intestine (De Fombelle *et al.*, 2001). This means that, despite corn containing more starch than oats on a weight for weight basis, if corn is fed whole, less actual starch will be digested in the small intestine and more will reach the hindgut. Feeding large amounts of corn starch is more likely to result in a significant decrease in hindgut pH and an increased risk of acidosis than if an equivalent amount of oat starch is fed.

There have been a number of studies that have evaluated the beneficial effect of cooking or micronising cereals such as corn and barley to improve pre-caecal digestibility (Kienzle 1994; Cuddeford 1999c; Mclean *et al.*, 2000). It also has been recommended that if meals high in starch are fed they should be fed separately from the fibre (Cuddeford 1999b). It is therefore worth considering increasing the number of meals/day to three or four rather than having two large cereal-based feeds. If a horse appears to require ever increasing amounts of feed in each of its meals in order to maintain condition and energy, and there is no suspicion of ill health consider

- Increasing the number of meals (whilst keeping down the size of each meal);
- Changing to a feed with a higher energy content;
- Adding oil.

6.3.1.5. *Feed changes.* Rapid or major changes in the amount or type of feed eaten by the horse may cause marked fluctuations in the microbial population of the hindgut. Although the microflora adapts to a degree to whatever diet is being fed, quick changes in their environment may cause gastrointestinal imbalances and may lead to diarrhoea or colic. In the past it has been considered to be much less of a risk with changes in preserved forage but as described above this may not always be true when changing to a less digestible source. Typically changing from one type of forage source to a similar one, causes fewer problems compared with changing from preserved to fresh forage or from a forage-only diet to one with concentrates, but ideally, any changes in the diet should be done gradually. Small changes can be made in a step-wise manner over 3–5 days, while more major changes may require a 2–3 week adaptation period. Changes in amount rather than type are preferable where possible and appropriate. Even in a grain-adapted animal the amount of grain should not be increased by more than 0.5 kg per day (for a 500 kg horse).

6.3.1.6. *Reducing the risk of sand colic.* To prevent sand colic, the horse must not be allowed to ingest any sand! Minimising sand ingestion in some geographical areas can be difficult, but there are a few precautions, which should be taken:

- Always ensure nutrition is optimal and the diet is not deficient.
- Do not graze on bare pastures. If this cannot be avoided then an ample amount of an alternative forage should be provided.
- If forage and feed are fed from the ground, this should preferably be a concrete area.
- Offer grass hay rather than alfalfa hay.
- Follow a regular anthelminthic programme.

For voracious sand ingestors, particularly those prone to colic, there are a number of psyllium or clay-based products which some believe assist in the removal of sand from the gut (Ruohoniemi *et al.*, 2001). However, opinions are divided, and work by Lieb and Weise (1999) found no advantage to sand removal from the GIT of feeding or treating with psyllium, wheat bran or oil at 1.5% BW hay intakes. They also reported that at larger hay intake (2.5% vs 1.5%) the sand moves through more rapidly, suggesting that maximising forage intake may be beneficial.

6.4. CHANGING BODYWEIGHT SAFELY

6.4.1. *Conditioning the underweight horse*
The ideal condition of any horse is dependent on its breed, work, life stage and health (for example, you would not wish to allow a horse with problems with its joints to become over-weight). The sick horse is outwith this chapter, however, there may be occasions when a healthy horse is below its desirable weight, and requires changes to its dietary regimen in order to improve its condition. Assuming that the horse has not been starved, has been on a suitable antihelminthic programme and has no dental problems, the rules for increasing weight are as outlined below. All changes must be made gradually, particularly to horses that are extremely under-weight, as rapid introduction of feed may result in death (Kronfeld 1993). The exact protocol to follow will depend on the extent of the starvation (Naylor 1999; Stull *et al.*, 2001) but in extreme cases should start with increasing the forage quantity and quality, ideally for at least 10 days, before introducing concentrates, which can be high fibre cubes to begin with. The horse hould be fed at the level to maintain its weight at the midpoint between its current and desired weight. Once this weight has been attained then the feed may be increased gradually to a level which will maintain the horse at the goal weight. Supplementary oil may be beneficially used in the re-feeding programme (Stull *et al.*, 2001). Such a programme of weight increase may take several weeks and care is needed with respect to mineral intake, in particular phosphorus and magnesium, especially in severe cases.

If the horse has trouble with maintaining weight (*i.e.* is a 'poor doer'), the forage should have a high digestible energy and additional energy in the form of oil rather than grain should be offered. Vitamin and mineral content of the diet should match the energy level, therefore if feeding below recommended levels of concentrates, or adding oil, it may be necessary to add a vitamin and mineral supplement. All of this is based on increasing the amount of digestible nutrients available to the horse. Changes may be required in other management factors such as for example, reducing heat loss through appropriate shelter/rugs.

6.4.2. *Reducing weight*

Ponies, especially pregnant ones, must not be abruptly starved to reduce their body weight or prevented from eating for prolonged periods as they have an increased risk of developing hyperlipaemia under such circumstances. It is much safer to gradually reduce the diet to a half maintenance level, if necessary, than to completely starve a pony for weight loss purposes. Wherever appropriate the diet can be made up to near appetite levels by feeding low energy forages but poorly digested, highly silicated forages such as wheat straw, may cause impaction.

6.5. GASTROINTESTINAL ULCERS

The horse secretes acid into the stomach continuously. Under natural circumstances the saliva produced by the horse whilst grazing, helps to buffer the gastric acid and may help to provide a protective layer for the squamous epithelium. In addition, living under natural conditions encourages the horse to move freely and may assist in the normal movement of stomach contents through the gastrointestinal tract. Unfortunately, modern management practices which include meal feeding, low fibre/high concentrate diets, early weaning and intensive training programmes, help to produce a poorly buffered, acidic environment in the stomach. This results in a high prevalence of gastrointestinal ulcers, particularly in intensively-managed horses, such as performance horses (Bertone 2000). The prevalence and severity of ulceration has been correlated with the intensity of training that the horse undergoes and the associated management practices. Murray *et al.* (1996) found that 93% of Thoroughbred racehorses had gastric ulcers, but this percentage varied with the training – two-year olds at the start of training showed no or minimal lesions, but after only 2–3 months of intensive training 90% showed lesions (Murray, unpublished data from Orsini, 2000). Contributing factors associated with lesions are high concentrate diets, low hay diets, meal feeding, fasting, training (which increases gastrin levels) and certain drugs. In addition, the stressful lifestyle of such performance horses is thought to exacerbate these conditions. By comparison, horses that undergo less intensive or no training, or even a different type of training, show a far lower prevalence, *e.g.* 37% in pleasure horses (Murray *et al.*, 1989), and up to 40% in Western performance Quarter Horses (Bertone 2000). GIT ulceration is not isolated to adults, but is prevalent in foals, with up to 51% of foals aged less than 3 months showing lesions (Murray *et al.*, 1989). Foals are highly susceptible to ulceration because they start secreting gastrin just after birth before the gastric mucosa have fully developed (Sandin *et al.*, 2000). In addition, stressful weaning programmes may exacerbate the development of GIT ulcers. Clinical signs of gastrointestinal ulcers include abdominal discomfort, reduced appetite, weight and body condition loss, diarrhoea, and particularly in foals, a loss of vitality, dorsal recumbency, grinding of teeth. The onset of crib-biting may indicate the presence of gastric lesions (Nicol *et al.*, in press).

6.5.1. *Risk reduction*

Preventing the onset of gastrointestinal ulcers should be a priority, and this can usually be achieved by keeping adult horses at pasture. However, as this is not possible for all horses, other steps often need to be taken. There is a strong correlation between the diet fed and the pH of the stomach. Concentrate diets have always been implicated but ensuring these are fed in small amounts, possibly in combination with a forages such as alfalfa hay may help to increase the stomach pH and volatile fatty acid content in gastric contents (Nadeau *et al.*, 2000). A high forage intake that encourages chewing and stimulates salivation should be maintained. If lesions do occur, training should be reduced or stopped until the lesions have healed. Wherever possible and especially in a horse with known predisposition to this problem, stressful situations such as travelling long distances, changing environment and long periods of confinement where the horse cannot freely move around should be avoided. There are a number of pharmaceutical agents that can be used in consultation with the veterinarian. These work by preventing or reducing the secretion of gastric acid and therefore increasing gastric pH, coating the mucosa or introducing endogenous prostaglandins, but they should ideally be viewed as a short-term measure to resolve the lesions while changes in management practices take effect. Omeprazole paste may be the most safe and effective anti-ulcer therapy to use in horses which continue to be trained and raced (Johnson *et al.*, 2001).

6.6. LAMINITIS

Laminitis has been recognised as a clinical disorder since the 19th century. It is currently thought to represent an end stage condition, common to a number of disease processes, which results in destruction of the interlaminar bonds of the hoof. If severe this may allow independent movement of the pedal bone with respect to the hoof wall (Howarth 1992). The laminae consist of interlocking finger-like projections that are separated by nerves and blood vessels (see Figure 1). It is these blood vessels that supply the hoof. The laminae hold an integral part in the construction of the hoof wall by providing strength and stability to the structure as a whole. The force created by the horse's weight is passed down through the bones of the leg to the top of the pedal bone (3rd Phalanx or P3), which is locked within the hoof wall. The laminae effectively connect the outside of the pedal bone to the inside of the hoof wall, transferring the force from one to the other, thus, distributing the horse's bodyweight over a much larger surface area.

Laminitis can affect any horse but there appears to be an increased incidence in overweight animals, small to medium size horses and in particular ponies. There are many potential causes. Certain episodes result as a consequence of a primary disease or condition elsewhere in the body, for example retained placenta, an endocrine imbalance (*e.g.* Cushings) or certain toxaemias.

Ralston (1992) stated that 'overall the reliance on oropharyngeal stimuli, lack of attention to gastrointestinal or metabolic feedback during the course of a meal in conjunction with relatively slow adaptation to changes in caloric density of feed, make horses prone to excessive intake (gastrointestinal overload) when suddenly

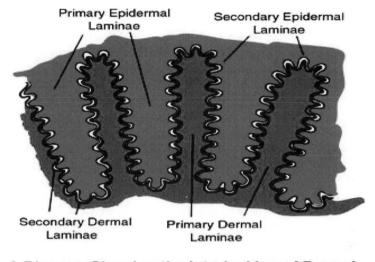

A Diagram Showing the Interlocking of Dermal and Epidermal Laminae within the Equine Foot

Figure 1. Laminae.

presented with feeds of high caloric density'. Grain overload whether by accident or deliberately induced increases the risk of developing laminitis. For over 50 years it has been known that the administration of a 85% carbohydrate and 15% fibre diet will result in clinical signs of laminitis within 40 hrs, including a pounding digital pulse, tachycardia, pain and a characteristic forelimb lameness (Obel 1948).

Turning ponies out onto lush pastures in the spring and autumn is a common triggering factor for the development of laminitis. Currently it is thought that the high levels of water soluble carbohydrates (which include the simple sugars as well as the more complex storage carbohydrate (fructans) may be involved in this process. It is thought that as for other mammals the horse does not have the necessary enzymes to digest fructans directly within the small intestine. They therefore pass into the hindgut where they are readily fermented in a similar manner to starch that escapes digestion (Cuddeford 1999a). Rapid fermentation will lead to production of lactic acid, and the lowering of pH (acidosis) upsetting the bacterial/microbial balance. When this occurs, bacteria that are not able to survive under such conditions may die and release endotoxins and other unwanted compounds into the hindgut. These endotoxins together with the other unwanted compounds could be absorbed into the blood and have further effects. The blood flow to the feet, is particularly sensitive to some of these factors and results in an inflammatory response, triggering the development of laminitis.

6.6.1. *Nutritional risk factors for laminitis*

Fructan (and water-soluble carbohydrate – WSC) intake should be reduced. The fructan content of grass will vary and is largely dependent on, light intensity, ambient temperature, stage of growth, residual fructan accumulation from the previous day and pasture management regimens (Cuddeford 1999a). It is usually when energy demands upon the plant are high (*e.g.* during growth) that the fructan concentration will be at it's lowest (as it is being utilised to provide energy). Fructan levels are higher in the stem than the leaf as it is a storage carbohydrate (Cuddeford 1999a). It is also known that particular grasses contain a higher concentration than others, for example, cocksfoot and timothy both have a lower level than most perennial ryegrasses. The fructan content of grass is likely to be higher during spring before the development of the flower when the fructans are being stored. In the autumn time, if a low temperature is followed by a bright and warm day the fructan concentration may also increase as the factors mentioned above may not be sufficient to support growth. It is therefore thought that the safest time to turn out may be late at night, bringing the horse in by mid-morning. The fructan content of hay is likely to be lower than that of fresh grass, and will be lower still in haylage due to the fermentation process. There are various practices that are recommended that will reduce the risk of the development of laminitis. These are summarised in Table 1.

Table 1. Recommended practices for reducing the risk of laminitis.

- Horses should only be at pasture when fructan levels are likely to be at the lowest, such as late at night to early morning, removing them from the pasture by mid-morning.
- Horses should not be grazed on pasture that has not been properly managed by regular grazing or cutting. (young leafy sward should be maintained at approximately 4 cm in height and mature stemmy grasses should be avoided since this contains high levels of stored fructans).
- Horses should not be turned out during the spring (before flower development) and autumn months when fructan levels are accumulating and plants generally contain relatively high levels of WSC.
- Horses should not graze pastures that have been exposed to low temperatures (e.g. frosts) with warm, bright sunny weather.
- Horses should not be grazed on freshly cut stubble (after hay has been harvested), as plant stems are the storage organs for fructans.
- For some horses it may be necessary to consider zero grazing (whilst providing the horse with suitable forage).
- If horses are grazed, they should only feed on grasses that produce low levels of fructan (eg. timothy).
- Horses should be fed forage rather than concentrates. Oil could be used in place of starch as an energy source (with veterinary approval) and broad spectrum vitamin and mineral supplements can be used where no (or low levels of) concentrates are fed and if the quality of the forage is poor.
- Horses should be fed according to their energy requirements.
- Horses should have their hooves regularly trimmed or shod.

6.7. RESPIRATORY SYSTEM

Allergic respiratory disease in horses or Recurrent Airway Obstruction (RAO) is more frequently referred to as 'heaves' or Chronic Obstructive Pulmonary Disease (COPD; Lavoie 1997) and is one of the oldest documented diseases of the horse. There is considerable evidence implicating allergic reactions in most affected animals. It is rarely found in tropical or subtropical countries where animals spend most of their time at pasture. In temperate climates COPD is closely associated with housing and the presence of dust and fungal spores in the air (Halliwell et al., 1993). The pathophysiology may be multifactorial, with a lower airways inflammation caused by a hypersensitivity reaction to specific allergens; a non-specific hyper-reactivity reaction or an inflammatory response induced by dust. Two moulds: *A. fumigatus* and *F. rectivirgula* have been most commonly implicated as inciting allergens (Halliwell *et al.*, 1993). COPD can affect up to 50% of horses in some stables resulting in a range of signs from reduced exercise-tolerance to, coughing, mucus hyper-secretion and severe dyspnoea. COPD can affect all types of horses and can be difficult to control especially if the animal is in full work (Andrews & Schemeitzel 1999; Robinson 2001). Some horses develop a similar condition when at pasture (McGorum & Dixon 1999) with no access to hay or straw – Summer Pasture Associated Obstructive Pulmonary Disease (SPAOPD). Some of these will develop classical COPD when exposed to poorly conserved hay/straw. It has been suggested (McGorum & Dixon 1999) that horses with SPAOPD have 'long term airway hyper-responsiveness which results in persistences of inflammation and dysfunction in response to a wide range of specific (*e.g.* mould or pollen allergens) and non specific triggers (*e.g.* cold air, dry air, exercise, irritant dusts).

A number of studies have evaluated the effects of environmental conditions during outbreaks of respiratory disease and one suggested that 'whereas healthy horses may be able to tolerate considerable variations in environmental conditions the presence of dust and other environmental contaminants can substantially extend the convalescent period once horses are affected with respiratory disease' (Burrell *et al.*, 1996). These workers found in their study that those horses housed on straw in loose boxes were twice as likely to suffer from lower airway disease as those kept on shredded paper in American barns.

An additional challenge to the respiratory system may occur for stabled horses due to raised ammonia levels in the stable. Stable hygiene practices and bedding choice will obviously have an effect on this. An additional factor may be diets that provide a higher protein intake than required, as excess protein cannot be stored and the excess nitrogen is removed primarily via urea in the urine. Higher protein intakes result in higher plasma urea concentrations, an increased water requirement and more urea excreted in urine. This urea is then converted by bacteria within the stable environment to ammonia. Lower respiratory tract inflammation can result from exposure to ammonia (Clarke 1987).

Hay quality can be very variable due to unpredictable weather conditions; hay often tends to be dusty with a high fungal spore count, especially the small diameter thermophilic actinomycetes which have been particularly associated with the dust from hay. It is not always possible to confirm the hygienic quality by eye (Raymond *et al.*, 1994). It has been estimated that around 12 million particles may be taken into the lungs of an average stabled horse with each breath and that even in well ventilated buildings the level of airborne particles around the hay and therefore within the breathing zone will still be very high (Blackman & Moore-Colyer 1998). Whilst obviously dusty and mouldy feeds should be avoided for all horses (but especially if they are prone to COPD or SPAOPD) other measures are often needed. There has been an increase in the use of alternative forage sources for horses, in particular big bale haylage (DM > 40%) and silages (< 40% DM). These, if prepared and stored correctly tend to have a better hygienic quality but there may be other concerns, especially with respect to potential clostridial activity and reduced overall fibre intake (in horses).

It has been reported that soaking a 2.5 kg hay net for 30 minutes reduces respirable particle numbers by 88% (Blackman & Moore-Colyer 1998). The practice of soaking hay with the aim of reducing the number of airborne particles is popular in some areas (Blackman & Moore-Colyer 1998; Harris 2000). The water used for soaking must be fresh and, although there is still discussion over exactly how long soaking should be carried out, increasingly it is recommended that more than 30 minutes may not be advisable because of the potential negative effect of prolonged soaking on the soluble carbohydrate and nitrogenous content of hay (Warr & Petch 1992; Blackman & Moore-Colyer 1998). Alternative solutions include the practice of steaming hay. In one study soaking for 10 or 20 minutes reduced the respirable particle numbers but also the concentration of phosphorus, potassium, magnesium, sodium and copper (Blackman & Moore-Colyer 1998). Steaming had a similar effect on the respirable particles but there was no loss of nutrients. However, these authors noted that, although reducing the number of respirable particles may be advantageous to a healthy horse and may help prevent the initiation of this debilitating condition, it may not be sufficient for some sufferers of COPD, in which case vacuum-packed dust-free fodder and pasture turnout would be recommended.

Recent research (Deaton *et al.*, 2001; Marlin *et al.*, 2001) found that ascorbic acid (vitamin C) levels were lower in the lung epithelial lining fluid and plasma of horses with COPD/RAO, compared to those of healthy horses. Further work (ongoing) suggested that lung function of RAO horses in remission could be improved by feeding a cocktail of antioxidants (Kirschvink 2001).

6.7.1. *Management of respiratory diseases*
There are a number of steps that can be taken to improve the environment for horses with RAO which may help to alleviate the condition (Table 2). However the first step should be to gain veterinary diagnosis and advice.

6.8. DEVELOPMENTAL ORTHOPAEDIC DISEASE (DOD)

The term DOD was first coined in 1986 to encompass all orthopaedic problems seen in the growing horse. It is non-specific and currently is taken to include osteochondrosis; subchondral cystic lesions; physitis; acquired angular limb

Table 2. Management to alleviate respiratory problems.

- The affected horse should be turned out as often as possible.
- The stable environment should be well ventilated (as opposed to draughty).
- Deep-litter bedding systems should be avoided.
- Faeces and urine should be removed frequently, and the horse should be moved from the stable before cleaning the bedding or grooming.
- The bedding substrate should provide a reduced allergen challenge and be non-dusty.
- Affected horses should not be housed near to any concentrated source of irritants/allergens such as, deep- litter beds, dung-piles, hay and straw stores.
- Hay should be checked and low dust hay should be fed. Hygienic quality should be checked if possible.
- If hay is fed, it should be soaked as this promotes the swelling of fungal spores so that they may be ingested rather than inhaled. The current recommendation is to soak hay for 30 minutes in clean water. Alternatively, steamed hay or other forage types such as haylage could be fed.
- All other feedstuffs should be of good quality.
- Feeding an appropriate antioxidant supplement formulated for horses should be considered.

deformities; flexural deformities; cuboidal bone malformation; juvenile arthritis or juvenile degenerative joint disease and bony fragments of the palmer/plantar surface of the first phalanx of standardbred horses (believed traumatic lesions) (McIlwraith 2001). Many possible causes have been suggested but in future it may be necessary to differentiate the cause and the triggering factor(s) from any consequent clinical signs. It has been suggested that as far as osteochondrosis is concerned it is most likely that initially some disturbance to the normal cartilage development occurs and then physical stresses are superimposed leading to the clinical signs of Osteochondrosis Dissecans (OCD) (McIlwraith 2001). It is also thought possible that the initial defects/lesions may heal or may develop into OCD or into subchondral bone cysts.

Many causes have been suggested for DOD, including a genetic disposition, biomechanical trauma, mechanical stress through inappropriate exercise, obesity, rapid growth and inappropriate or imbalanced nutrition. It is currently thought to be multi-factorial in origin. With respect to nutrition various nutrients have been implicated over the years but in particular protein, energy, calcium, phosphorus, copper and zinc (Kronfeld *et al.*, 1990; McIlwraith 2001).

There is currently little evidence that protein *per se* is a major factor but too much or the wrong type of energy may be involved in some cases. Feeding 129% NRC energy requirements to foals from 130 days of age resulted in increased incidence of lesions compared with controls or horses fed 126% National Research council's recommendations (NRC 1989) for protein (Savage *et al.*, 1993). It has been suggested that diets or individuals that respond to diets to produce high glyceamic peaks, which in turn cause high insulin peaks and the consequent reduction in the production of thyroxine which affects bone maturation, may have an increased risk of developing DOD (Glade & Belling 1986; Ralston 1995; Pagan 2001). Savage *et al.* (1993) fed three and a half times NRC recommended levels of calcium to 6 foals and found no change in incidence. However, although the incidence of clinical physitis and flexural deformities has not been shown to be related to marginal deficient phosphorus levels, excessive phosphorus intakes appeared to increase the incidence of OCD in unexercised foals – but these intakes were being fed with concurrent high energy intakes (129% NRC) (Cymbulak & Christison 1989).

A copper containing enzyme – Lysly oxidase, is involved in the X-linking of protein chains in elastin and collagen of cartilage. Disruption of these X-links due to copper deficiency may result in biomechanically weakened cartilage and increase the risk of DOD (Hurtig *et al.*, 1993). A number of studies have suggested a relationship between copper and zinc in the diet and DOD (Knight *et al.*, 1985, McIlwraith 2001) but whilst improved diets may have helped some horses, this does not appear to have been universally successful and there may have been an effect of study design on the results. In a series of elegant studies in New Zealand (Pearce *et al.*, 1998) copper supplementation of mares significantly decreased the radiographic indices for physitis in the distal third metatarsal bone of foals at 150 days and the prevalence of articular cartilage lesions. Supplementing the mare did increase the liver copper status but had no effect on other tissue copper concentrations. It is important to note that, unlike many of the US based studies, articular cartilage lesions were minor in all foals with no evidence of clinical OCD *in vivo*, with the exception of minor radiographic changes assessed at post mortem. The differences in the results may reflect other nutritional factors and differences between the US and pastured foals in New Zealand. Copper supplementation of the foal did not have any effect on any of the bone or cartilage parameters measured but the copper intake of the control foals was only mildly deficient.

Excessive zinc has been associated directly with OCD – suggesting that increased exposure to zinc and possibly cadmium may result in the development of OCD.

6.8.1. *Reducing the risk of DOD*

Adequate and balanced nutrition with appropriate exercise for the growing foal is very important and although this will not totally reduce the risk of DOD it might help reduce it. This is paramount as once DOD is evident, the solutions, such as early weaning or confinement, while perhaps being effective at reducing the impact of the disease, severely reduce other welfare aspects of the foal's life.

Steady growth curves should be aimed for, and any spurts are to be avoided. Diets should be nutritionally balanced, preferably avoiding those with high-energy intakes from hydrolysable carbohydrates. It may also be advisable to avoid high protein diets that also provoke high insulin responses. Kronfeld's group at Virginia Tech, USA in particular, have been evaluating the benefits of feeding a fat and fibre rich pasture supplement compared with a more traditional starch and sugar based supplement and have found:

- Smoother growth curves without the spring 'dip' and compensatory surge
- Reduced glycaemic response – more equivalent to that seen at grass
- Less disturbance to the somatatrophic axis (see Hoffman *et al.*, 1999; Stanier *et al.*, 2001; Williams *et al.*, 2001).

Unless severe lesions already exist, exercise may help to modulate the adverse effects of high-energy diets on joint development. Bruin *et al.* (1994) exercised warm-blood foals fed high energy diets (30–45 mins of trot and gallop in a lunging ring several times a day) and found beneficial effects. However, in a recent study exercise was not shown to reduce the number of lesions but it was suggested that it might help reduce the severity (Van Weeren & Barneveld 1999). In conclusion, it is currently recommended that adequate exercise should be provided daily. Excessive strain or trauma to the young horses' growing bones and joints should be avoided. This can be done by avoiding exercise to fatigue/exhaustion and also by avoiding a situation of confinement followed by abrupt turn out, especially on hard ground, as this increases the risk of biomechanical trauma.

Adequate calcium in balance with phosphorus is advisable for the growing animal. Certainly excessive calcium may not be protective and may interfere with the absorption of other elements. Pregnant mares should be fed adequate copper and zinc to allow the foetus to accumulate sufficient stores; therefore, growing foals and pregnant mares should be fed approximately 15–20 mg copper/kg DM. Zinc should be fed at between 3–5:1 (Zn:Cu); optimally at 4:1 Zn:Cu.

7. Nutritional Impact on Behaviour

An inappropriate diet, whether it is a deficiency or an overdose of a specific nutrient or feed type, can be reflected in changes in behaviour. Behaviour which may be considered by an owner to be inappropriate, such as pica/geophagia (McGreevy *et al.*, 2001), wood chewing or copraphagia may be the result of a deficient diet (although in young horses it may also reflect their motivation to investigate and orally manipulate all kinds of substrates).

Obviously, any dietary factor that causes the horse pain, such as rapid or large changes in the amount of concentrates fed, will result in the horse responding differently to its normal environment. Hind-gut acidosis or other gastrointestinal disturbances may not result in the visible symptoms of full blown colic, but the consequences may be apparent as more subtle behavioural changes. The horse may become depressed, cantankerous, show increased sensitivity to touch or just show changes in responsiveness. Changes in activity levels and type may also indicate pain, such as moving less, standing in a manner that suggests tension rather than a relaxed state, or alternatively the horse may become more restless, or display specific behavioural changes, such as constantly turning around to nip its side.

There are many behavioural changes that are recognised symptoms of specific diet deficiencies or toxicities which are too extensive to be covered here; in addition, some behavioural changes may not be linked to diet but to a specific condition or disease. However, it is becoming increasingly recognised that some specific behaviour patterns that once were thought to be due to boredom and were considered to be vices, may be linked to modern management practices, and specifically to diet (although stable confinement is also a major factor). These behaviour patterns tend to be classified as stereotypies and include both oral and locomotory behaviours (see Chapters 4 and 5). Time spent foraging, and also a reduction of appropriate forage given concomitantly with an increase in concentrates fed has been indubitably linked to displays of abnormal behaviour patterns. For instance, crib-biting – an oral stereotypy not recorded in wild horses – was recently studied in young foals (Nicol *et al.*, 2001), and a link was established between the condition of the stomach, specifically the presence of changes in gastric morphology and gastric lesions, of the foals and crib-biting.

8. Nutritional Supplements

Nutritional supplements are readily available to horse owners, both on prescription and from retailers and come in many forms: powders, pastes, drenches, herbs etc. The vast majority are widely used without the recommendation from either a veterinarian or a nutritionist. Nutritional supplements are fed for many reasons, and the benefits they proffer include the following:

- an additional source of vitamins and minerals;
- 'diet enrichment' through providing plants that horses may not have access to;
- ergogenic aids (which claim to enhance performance);
- health benefits (*e.g.* which claim increased immunity, improved hoof appearance, better respiratory function);
- behavioural modification (*e.g.* for 'hormonally challenged' mares or excitable horses).

There are a number of issues that should be taken into account when considering nutritional supplements, but the key ones are: how necessary, how efficacious and how safe is the supplement in question. Some supplements that contain active ingredients have been clinically tested in horses and have proof of efficacy and safety, others may not have been clinically tested in horses and have no published proof of efficacy. There are even supplements being sold that contain ingredients that are known not to be absorbed or efficacious in the horse. There are many supplements available, which claim unsubstantiated benefits, and it would be preferable, if perhaps rather optimistic, that all products should be clinically tested with results published in peer reviewed journals. It is often the case that easy solutions (or so called 'quick fixes') are sought for behavioural problems or lack of training that would be better solved with some time spent working with the animal.

To assist in the question 'to supplement or not?' the following areas should be considered. Firstly, before offering any horse a supplement outwith its normal diet, there are a number of questions that should be asked:

- Is the horse being provided with the optimum balanced diet for their current health and lifestage? Manufacturers of feedstuffs that are intended to provide the concentrate portion of a horse's diet will have the produced a mineral and vitamin level that is designed for a horse in a certain lifestage and/or under a certain workload, i.e they will be in ratio with the calorific content. If the horse does not require the recommended amount of that feed, and it is not possible to change over to a more suitable feedstuff then it may be necessary to supplement the horse's ration with additional minerals and vitamins, in order to supply all its' nutritional requirements.

- Is the horse being provided with the optimum environment – this includes ability to forage and interact with conspecifics. Sub-optimal condition or poor performance or behaviour may be a reflection of an inadequate environment, which may impact both on the horse's physiological and psychological well-being. Supplements may have no impact or may mask the true root of the problem.
- Is the horse capable of fulfilling the task required – *i.e.* is it being provided with the correct training (in saddle and groundwork) and exercise level to fulfill the task required? All athletes need to be properly trained and in peak fitness to perform their best, the horse is no different. Ergogenic aids should be considered only as one of the additional tools available, and should not be used as a replacement for appropriate training. In addition, many performance or behaviour issues can be resolved more permanently with patience, time and a regular modification/training programme, *i.e.* there really is no substitute for time spent working with the animal in question.
- Are there any health issues that require veterinary treatment that may need to be addressed? It may be neither ethical nor safe to use supplements to mask pain in order to get good performance from your horse. One of the purposes of pain can be to restrict movement within the physical bounds in order to minimise the risk of further damage. In addition, pain is often the cause of behavioural problems in horses (Casey 1999) and steps to ensure that the horse is pain-free and that any equipment being used is appropriate should be paramount.

If the nutrition, the environment, the training and the health issues have been addressed, and there is still a perceived requirement for improvement – whether it be performance, condition or behaviour, then supplements may be considered. At this point, the next set of questions should be about defining the problem – is it a general loss of condition or are there specific indicators about what may be most beneficial. For example, the supplements that are offered may differ for poor hoof or coat condition as opposed to a general loss of body condition. In addition, the advice of a nutritionist or a veterinarian with nutritional training will help with the decision making process. If anything is being added to the diet it is important to ensure that the addition to the horse's normal diet does not take any one nutrient over recommended levels in total *e.g.* selenium is often added to the diet of horses but it has been recommended that total daily intakes do not exceed around 1 mg per 100 kg bodyweight. In addition it is necessary consider the response time of the supplement, and the different responses of individuals, since some nutritional supplements are required to be fed for long periods before a noticeable response occurs, *e.g.* it is known that biotin needs to be fed for several months before any improvement to poor hoof quality may be visible in those individuals in which it is effective.

Once the problem has been defined, and the desired action or active ingredients, which are required, have been identified, how do you choose a safe and efficacious supplement? It cannot be taken for granted that all products are either efficacious or safe for horses, so care must be taken. There are supplements which have no proof of efficacy, have never been clinically tested on horses and for which no toxicity data in horses is available (see Poppenga 2001). Many product manufacturers claim that a ingredient 'X' is well known for causing response 'Y', but they fail to mention that all of the data is from human experiments, and although some extrapolations from human data to horses have been well founded, there are many which are not (Harris & Harris 1998). Often more intensive evaluation of supplements on the market can reveal inadequate experimental design and inaccurate assumptions. For example some vitamins and minerals can be toxic at high levels and the active ingredients in some herbs are a drug used by humans *e.g.* salycilic acid (aspirin) obtained from the bark of *Salix alba* (white willow). In this case more may not be better. Finally it is essential before using a supplement that contra-indications are assessed. There have been few satisfactory studies that cover this area, although some side-effects are known. For example it is advised that gingko is not used with aspirin, non steroidal anti-inflammatories or anti-coagulants due to its profound inhibition of platelet-activating factor (Davidson 1999).

9. Conclusions

The horse has evolved to fulfil a specific niche as a social animal, grazing and browsing throughout the day, conscious of, and ready to respond to, its' position as a prey item – and as such, is very successful in this role. The tasks required of today's horse, and the different management practices, including diet, imposed as a consequence of this role, do not always match its' physical or mental needs, and often may be the source of problems that develop either in the horse's well-being or in the relationship between horse and owner. While it is unrealistic to believe all horses will ever be maintained under wholly natural conditions (indeed their status in society could be drastically reduced if that were the case), we should encourage people to alter their management practices, examine their feeding programmes and make changes where possible (Davidson 1999). In summary, the following points should be considered in order to prevent the onset of nutritionally related problems and enhance the welfare of the performance horse:

- Where possible 24 hour access to grazing should be allowed. Grazing quality should be appropriate – many modern pastures are rich monocultures with too little fibre content in the grass, and therefore may need to be supplemented or managed to provide more optimal grazing. Nutritional supplementation, especially vitamins and minerals, may be needed especially for the performance horse, pregnant/lactating mare and the growing animal.
- If stabled, as much of the diet as possible, should be forage, and preferably a selection of different forages should be offered. Recent work (Goodwin *et al.,* 2001) found that at least in the short term, stabled horses showed a preference

for a stable with multiple forages over a stable with the standard single forage. Providing different types of forages to the stabled horse reflects in a small way the more varied diet available to grazing horses.

- Where appropriate to the individual, ad-libitum hay should be provided. Although some horses initially gorge on ad-lib hay, after about 12 days of ad-lib hay, most horses will stabilise to a constant intake.
- If high-energy complementary feeds are required, they should be introduced gradually and fed in small amounts in multiple meals rather than in one large meal. As an alternative to feeding high grain diets, the horse could be maintained on a diet composed of fibre, with additional energy added in the form of oil (remembering to supplement with vitamins and minerals as necessary).
- Feed should be changed very gradually.
- The diet should be nutritionally balanced, either too little or too much of any nutrient (whether it be protein, a vitamin or carbohydrate) compared to the individual horse's particular nutritional requirements, may cause changes in behaviour or clinical problems (such as colic).
- Plants that are poisonous to horses should be identified and horses should not have access to them.
- The horse should live a stress-free life. Stress may come in many guises and may be individual to that horse – travelling, removing stable companions, weaning, or increased intensity in training. Any changes in the management practice of the individual horse and its environment, should be minimised and introduced gradually.

10. References

Andrews, F. and Schemeitzel, L. (1999) An update on Chronic obstructive pulmonary disease in horses. *Veterinary Medicine*, 171–181.

Anon-BETA (1999) Survey undertaken by the Produce Studies Group on behalf of *British Equestrian Trade Association*, West Yorkshire, UK.

Archer, M. (1971) Preliminary studies on the palatability of grasses, legumes and herbs to horses. *Veterinary Record* **89**, 236–240.

Archer, M. (1973) The species preferences of grazing horses. *J. British Grassland Society* **28**, 123–128.

Archer, M. (1978) Further studies on palatability of grasses to horses. *J. British Grassland Society* **33**, 239–243.

Avery, A. (1996) *Pastures For Horses, A Winning Resource*. Rural Industries Research and Development Corporation Gillingham printers Adelaide, Australia.

Bertone, J.J. (2000) Prevalence of gastric ulcers in elite, heavy use Western performance horses. *J. Veterinary Internal Medicine* **14**, 366 (abstract).

Blackman, M. and Moore-Colyer, M.J.S. (1998) Hay for horses: the effects of three different wetting treatments on dust and nutrient content. *Animal Science* **66**, 45–750.

Brown, J.H. and Powell Smith, V. (1994) *Horse and Stable Management*. Blackwell, London 16, UK.

Bruin, G., Creemers, J.J.H.M. and Smolders, E.E.A. (1992) Effect of exercise on osteochondrosis in the horse. *Equine Osteochondrosis in the '90s University of Cambridge*, 41–42 (abstract).

Burrell, M.H., Wood, J.L.N., Whitwell, K.E., Chanter, N., Mackintosh, M.E. and Mumford, J.A. (1996) Respiratory disease in thoroughbred horses in training: the relationships between disease and viruses, bacteria and environment. *Veterinary Record* **139**, 308–313.

Casey, R.A. (1999) Recognising the importance of pain in the diagnosis of equine behaviour problems. In Harris, P.A., Gomarsall, G., Davidson H.P.B. and Green R. (eds.), *Proceedings of the British Equine Veterinary Association Specialist Meeting on Nutrition and Behaviour*, pp. 25–28.

Carson, K. and Wood-gush, D.G.M. (1983) Equine Behaviour 11 Review of the literature on feeding, eliminative and resting behaviour. *Applied Animal Ethology* **10**, 179–190.

Clarke, A. F. (1987) Stable environment in relation to the control of respiratory diseases, in Hickman J (ed) Horse Management Academic Press, London, pp. 125 –174.

Clarke, L.L., Roberts, M.C. and Argenzio, R.A. (1990) Feeding and digestive problems in horses. Physiologic responses to a concentrated meal. *Vet Clinics of N. America: Equine Practice* **6, 2**, 433–450.

Cohen, D., Gibbs, P.G. and Woods, A. M. (1999) dietary and other management factors associated with colic in horses. *J. American Veterinary Medicine Association* **215**, 53–60.

Cuddeford, D. (1999a) Sugar is bad for my horses isn't it? In Harris, P.A., Gomarsall, G., Davidson, H.P.B. and Green R. (eds.), *Proceedings of the British Equine Veterinary Association Specialist Meeting on Nutrition and Behaviour*, pp. 69–72.

Cuddeford, D. (1999b) Why feed fibre to the performance horse. In Harris, P.A., Gomarsall, G., Davidson, H.P.B. and Green, R. (eds), *Proceedings of the British Equine Veterinary Association Specialist Meeting on Nutrition and Behaviour*, pp. 50–54.

Cuddeford, D. (1999c) Recent advances in equine nutrition. *Recent Advances in Animal Nutrition in Australia* **12**, 99–105.

Cymbulak, N.F. and Christison, G.I. (1989) Effects of dietary energy and phosphorus content on blood chemistry and development of growing foals. *Equine Veterinary J.* **S16**, 1993–1926.

Davidson, H.P.B. (1999) Natural horse – unnatural behaviour: why understanding natural horse behaviour is important. In Harris, P.A., Gomarsall, G., Davidson H.P.B. and Green R. (eds.), *Proceedings of The British Equine Veterinary Association Specialist Meeting on Nutrition and Behaviour EVJ*, pp. 7–10.

Davidson, H.P.B. (1999) Herbs – a sage of all wisdom or a waste of thyme? In Harris, P.A., Gomarsall, G., Davidson, H.p.b. and Green R. (eds.), *Proceedings of the British Equine Veterinary Association Specialist meeting on Nutrition and Behaviour EVJ*, pp. 32–34

Deaton, C., Marlin, D.J., Smith, N., Roberts, C., Kelly, F., Harris, P and Schroter, R.C. (2001) Systemic and Pulmonary Bioavailability of Two Different Forms of Ascorbic Acid in Equids. *Proceeding Waltham International Symposium*, p. 88.

De Fombelle, A., Frumholtz, P., Poillion, D., Drogoul, C., Phillipeau, C., Jacotot, E. and Julliand, V. (2001) Effect of Botanical Origin of Starch on Its Prececal Digestibility Measured with the Mobile Bag Technique. *Proceeding of Equine Nutrition and Physiology Society*, pp. 153–155.

Dixon, P.M. (2000) Removal of dental overgrowths. *Equine Veterinary Education* **12**, 84–91.

Feige, K., Schwatzwald, C., Furst, A., and Kaser-Hotz, B. (2000) Esopageal obstruction in horses: as retrospective study of 24 cases. *Canadian Veterinary J.* **41**, 207–210.

Frape, D.L. (1998) *Equine Nutrition and Feeding*. Blackwell Science Ltd., Oxford, UK.

Gill, E.L. (1988) *Factors Affecting Body Condition of New Forest Ponies*. PhD thesis, Department of Biology, University of Southampton, UK.

Glade, M.J and Belling, T.H. (1986) A dietary etiology for osteochondritic cartilage. *J. Equine Veterinary Science* **6**, 151–155.

Goodwin, D., Davidson, H.P.B. and Harris, P.A. (in press) Foraging enrichment for stabled horses: effects on behaviour and selection. *Equine Veterinary J.*

Halliwell, R.E.W., McGorum, B.C., Irving, P. and Dixon, P. (1993) Local and systemic antibody production in horses affected with chronic obstructive pulmonary disease. *Veterinary Immunology Immunopathology* **38**, 201–215.

Hansen, R.M. (1976) Foods of free roaming horses in Southern New Mexico. *J. Range Management* **29, 4**, 437.

Harkins, D., Smith, R.A. and Tobin, T. (1997) Poisonous plants and feed related poisonings. In *The Veterinarians Practical Reference to Equine Nutrition*. Purina Mills and American Association of Equine Practitioners, USA.

Harris, P.A. (1997a) Energy requirements of the exercising horse. *Annual Review of Nutrition* **17**, 185–210.

Harris, P.A. (1997b) Feeds and Feeding in the United Kingdom. In Robinson, N.E. (ed.), *Current Therapy In Equine Medicine* 4. WB Saunders, UK, pp. 698–703.

Harris, P.A. (1999a) How understanding the digestive process can help minimise digestive disturbances due to diet and feeding practices. In Harris, P.A., Gomarsall, G., Davidson, H.P.B. and Green, R. (eds.), *Proceedings of the British Equine Veterinary Association Specialist Meeting on Nutrition and Behaviour*, pp. 45–50.

Harris, P.A. (1999b) Feeding and management advice for Tying up. In Harris, P.A., Gomarsall, G., Davidson, H.P.B. and Green, R. (eds.), *Proceedings of the British Equine Veterinary Association Specialist Meeting on Nutrition and Behaviour*, pp. 100–104.

Harris, P.A. (2000) Feeding Practices in the UK and Germany. *Proceedings for the 2000 Equine Nutrition Conference for Feed Manufacturers*. Kentucky Equine Research Inc., pp. 241–267.

Harris, P.A. and Harris, R.C. (1998) Nutritional ergogenic aids in the horse – uses and abuses. In Lindner, A. (ed.), *Proceedings of the Conference on Equine Sports Medicine and Science*. Waageningen Press, The Netherlands, pp. 203–218.

Hastie, J.M.D. and Longland, A.C. (2001) In vitro fermentation of high temperature dried alfalfa and sugar beet pulp. *Equine Nutrition Physiology Society*, 32.

Hintz, H.F. (1994) Nutrition and colic. *Equine Practice* **16, 10**, 10–15.

Hintz, H.F. (1997) Hay racks vs. feeding hay on the stall floor. *Equine Practice* **19**, 5–6.

Hoffman, R., Lawrence, L.A., Kronfeld, D.S., Cooper, W.L., Sklan, D.L., Dascanio, J.J. and Harris, P. (1999) Dietary Carbohydrate and fat influence radiographic Bone mineral Content of Growing foals. *J. Animal Science* **77**, 3330–3338.

Howarth, S. (1992) Laminitis – an end stage endocrinopathy. *Equine Veterinary Education* **4**, 123–126.

Hurtig, M.B., Mikuna-Tagagaki, Y. and Choi, J. (1992) Biochemical evidence for defective cartilage and bone growth in foals fed a low copper diet. *Equine Veterinary J.* **S16**, 66–73.

Hyslop, J.J., Thomlinson, A. L., Bayley, A. and Cuddeford, D. (1998) Development of the mobile bag technique to study the degradation dynamics of forage feed constituents in the whole digestive tract of equids. *Proceedings of the British Society of Animal Science*, p. 129.

Hyslop, J.J. (1998) Modelling Digestion in the Horse. *Proceedings of an Equine Nutrition Workshop*. HBLB London, UK.

Kienzle, E., Radicke, S., Wilke, S., Landes, E. and Meyer, H. (1992) Praeileale Starke verdauung in Abhangigkeit von Starkeart und-zubereitang (Pre-ileal starch digestion in relation to source and preparation of starch), II. *Europische Konferenzieber die Ernhrung des Pferdes*, Hannover, pp. 103–106.

Kienzle, E. (1994) Small intestinal digestion of starch in the horse. *Revue De Medecine Veterinaire* **145, 2**, 199–204.

King, C. (1999) *Preventing Colic in Horses*. Paper Horse North Carolina, USA.

Kirschvink, N., Fievez, L., Bounet,V., Art, T., Degand, G., Smith, N., Marlin, D., Roberts, C., Harris, P. and Lekeux, P. (in press) Effect of nutritional antioxidant supplementation on systemic and pulmonary antioxidant status, airway inflammation and lung function in heaves-affected horses. *Equine Veterinary J.*

Knight, A.P. and Walter, R.G. (2001) *A Guide to Plant Poisoning: Of Animals in North America*. Teton New Media Jackson, WY, USA.

Knight, D.A., Gabel, A.A., Reed, S.M., Embertson, P.M., Bramlage, L.R. and Tyznik, W.J. (1985) Correlation of dietary minerals to incidence and severity of metabolic bone disease in Ohio and Kentucky in *Proceedings 31st Annual Meeting American Association of Equine Practioners*, pp. 445–561.

Kronfeld, D.S., Meacham, T.N. and Donoghue, S. (1990) Dietary aspects of developmental orthopedic disease in young horses. *Veterinary Clinics of North America: Equine Practice* **6, 2**, 451–465.

Kronfeld, D.S. (1993) Starvation and malnutrition of horses: recognition and treatment. *J. Equine Veterinary Science* **13**, 298–304.

Kronfeld, D.S., Cooper, W.L., Greiwe-Crandell, K.M., Gay, L.A., Hoffman , R.M., Holland, J.L.,

Wilson, J.A., Sklan, D. and Harris, P.A. (1996) Supplementation of pasture for growth. *2nd European Conference On Horse Nutrition: Nutrition And Nutritional Related Disorders Of The Foal*, pp. 317–319.

Kronfeld, D.S. (1997) Nutritional assessment in equine practice, *The Veterinarians Practical Reference to equine Nutrition.* Purina Mills and American Association of Equine Practitioners 171–194.

Kronfeld, D.S. and Harris, P.A. (1997) Feeding the equine athlete for competition. *The Veterinarians Practical Reference to equine Nutrition.* Purina Mills and American Association of Equine Practitioners, pp. 61–79.

Kronfeld, D.S., Cooper, W.L., Griewe-Crandell, K.M., Gay, L.A., Hoffman, R.M., Holland, J.L.,Wilson, J.A., Sklan, D.J., Harris, P.A. and Tiegs, W. (1998) *Studies of Pasture Supplementation, Equine Nutrition Conference for Feed Manufacturers.* Kentucky Equine Research Inc., pp. 41–43.

Lavoie, J. (1997) Chronic Obstructive pulmonary disease. In Robinson, N.E. (ed.), *Current Equine Therapy in Equine Medicine 4.* WB Saunders Philadelphia, pp. 79–127.

Lewis, L.D. (1995) Equine Clinical Nutrition. *Feeding and Care.* Lea and Febiger, London, UK.

Lieb, S. and Weise, J. (1999) A group of experiments on the management of sand intake and removal in equine. *Proceedings of the 16th Equine Nutrition and Physiology Symposium*, p. 257.

Johnson, J.H., Vatistas, N., Castro, L., Fisher, T., Pipers, F.S. and Maye, D. (2001) Field survey of the prevalence of gastric ulcers in thoroughbred racehorses and on response to treatment of affected horses with omeprazole paste. *Equine Veterinary Education* 13, 221–224.

Marie, T. (1999) More than they can swallow. *Equus* 262, 33–40.

Mariner, S. (1980) *Selective Grazing Behaviour in Horses.* PhD thesis Univ of Natal Durban S.Africa.

Marlin, D.J. Deaton, C.D., Smith, N.C., Roberts, C.A., Kelly, F., Harris, P. and Schroter, R.C. (2001) Development of a Model of Acute, Resolving Pulmonary Oxidative Stress in the Horse by Ozone Exposure. *Proceedings of the World Equine Airways Symposium.* Edinburgh, p. 30.

McGorum, B. and Dixon, P.M. (1999) Summer pasture associated obstructive pulmonary disease 9SPAOPD): an update. *Equine Veterinary Education* 11, 121–123.

McGreevy, P.D., Hawson, L.A., Habermann, T.C. and Cattle, S.R. (2001) Geophagia in horses: a short note on 13 cases. *Applied Animal Behaviour Science* 70, 119–125.

McIlWraith, C.W. (2001) Developmental orthopaedic disease (DOD) in horses a multifactorial process. *Proceedings of the 17th Symposium of Equine Nutrition and Physiology Society*, pp. 2–23.

Mclean, B.M.L., Hyslop, J.J., Longland, A.C., Cuddeford, D. and Hollands, T. (2000) Physical processing of barley and its effects on intracaecal fermentation parameters in ponies. *Animal Feed Science and Technology* 85, 79–87.

Martin-Rosset, W., Vermorel, M., Doreau, M., Tisserand, J.L. and Andrieu, J. (1994) The French horse feed evaluation systems and recommended allowances for energy and protein. *Livestock Production Science* 40, 37–56.

Mellor, D.J., Love, S., Reeves, M.J., Gettinby, G. and Reid, S.W.J. (1997) A demographic approach to equine disease in the northern UK through a sentinel practice network. *Epidemiologie Sante Animaux*, 31–32.

Meyer, H., Ahlswede, L. and Reinhardt, H.J. (1975) Untersuchungen uber Frebdauer, Kaufrequenz und Futterzerkleinerung beim Pferd. *Deutsche Tierarzliche Wochenschrift* 82, 49–96.

Meszoly, J. (2001) Don't forget to float. *Equus* 287, 38–46.

Moore-Colyer, M., Hyslop, J.J., Longland, A.C. and Cuddeford, D. (1997) Degradation of four dietary fibre sources by ponies as measured by the mobile bag technique. *Proceedings of 15th Equine Nutrition and Physiology Symposium.* Texas, pp. 118–119.

Moore-Colyer, M.J.S. and Longland, A.C. (2001) The effect of plain sugar beet pulp on the in vitro gas production and in vivo apparent digestibilities of hay when offered to ponies. *Proceedings of 17th Equine Nutrition and Physiology Symposium*, pp. 145–147.

Murray, M.J., Schusser, G.F., Pipers, F.S. and Gross, S.J. (1996) Factors associated with gastric lesions in Thoroughbred horses. *Equine Veterinary* J. **28, 5**, 368–374.

Murray, M.J., Grodinsky, C., Anderson, C.W., Radue, P.F. and Schmidt, G.R. (1989) Gastric ulcers in horses: a comparison of endoscopic findings in horses with and without clinical signs. *Equine Veterinary* J. *Suppl.* 7, 68–72.

Nadeau, J.A., Andrews, F.M., Mathew, A.G., Argenzio, R.A., Blackford, J.T., Sohtell, M. and Saxton,

A.M. (2000) Evaluation of diet as a cause of gastric ulcers in horses. *American J. Veterinary Research* **61**, 7, 784–790.

Nash, D. (2001) Estimation of intake in pastured horses. *Proceedings of 17th Equine Nutrition and Physiology Symposium*, pp. 161–167.

Nash, D.G. and Thompson, B. (2001) Grazing behaviour of thoroughbred weanlings on temperate pastures. *Proceedings of 17th Equine Nutrition and Physiology Symposium*, pp. 326–327.

Naylor, J. M., Robinson, J. A. and Bertone, H.J. (1993) Familial incidence of hyperkalaemic periodic paralysis in quarter horses. *J. American Veterinary Medicine Association* **200**, 540.

Naylor, J. (1999) How and what to feed a thin horse with and without disease. In Harris, P.A., Gomarsall, G., Davidson, H.P.B. and Green, R. (eds.), *Proceedings of the British Equine Veterinary Association Specialist meeting on Nutrition and Behaviour*, pp. 81–86.

Nicol, C.J. (1999) Understanding equine stereotypies. *Equine Veterinary J. Supplement* **28**, 20–25.

Nicol, C.J., Davidson, H.P.B., Harris, P.A, Waters, A.J. and Wilson, A.D. (in press) Crib-biting is associated with mucosal inflammation and ulceration in young horses. *Equine Veterinary J.*

NRC (1989) *Nutrient Requirements of Horses*, 5th edition. National Academy Press, Washington DC, USA.

Obel, N. (1948) Studies on the histopathology of acute laminitis Almqvist and Wiksells Boktryckteri Ak., Uppsala.

Orsini, J. (2000) Gastric ulceration in the mature horse: a review. *Equine Veterinary Education* **12**, 1, 24–27.

Pagan, J.D., Harris, P.A., Kennedy, M.A.P., Davidson, N. and Hoekstra, K.E. (1999). Feed type and intake affect glycemic response in thoroughbred horses. *Proceedings of Equine Nutrition Conference for Feed Manufacturers, Kentucky Equine Research Inc.*, pp. 147–149.

Pagan, J.D. (1999) Energy and the performance horse. In Harris, P.A., Gomarsall, G., Davidson, H.P.B. and Green R. (eds.), *Proceedings of The British Equine Veterinary Association Specialist Meeting on Nutrition and Behaviour*, pp. 60–62.

Pagan, J.D. (2001) The relationship between glycaemic response and the incidence of OCD in thouroughbred weanlings a field study. *Proceedings of the 47th Annual Conference of American Association of Equine Practitioners*.

Pearce, S.G., Grace, N.D., Wichtel, J.J., Firth, E.C. and Fennessy, P.F. (1998) Effect of copper supplementation on copper status of pregnant mares and foals. *Equine Veterinary J.* **30**, 200–203.

Pilliner, S. (1996) *Horse Nutrition and Feeding*. Blackwell Science Ltd., Oxford, UK.

Potter, G.D., Arnold, F.F., Householder, D.D., Hansen, D.H. and Brown, K.M. (1992) *Digestion of Starch in the Small or Large Intestine of the Equine*. Europische Konferenzieber die Ernhrung des Pferdes Hannover, pp. 107–112.

Poppenga, R.H. (2001) Risks associated with the use of herbs and other dietary supplements. In Turner, S.A. and Galey, F.D. (eds.), *The Veterinary Clinics of North America, Equine Practice, Toxicology* **17**, 3, 455–477.

Proudman, C.J. (1991) A two year, prospective survey of equine colic in general practice. *Equine Veterinary J.* **24**, 90–93.

Proudman, C.J., French, N.P. and Trees, A.J. (1998) Tapeworm infection is a significant risk factor for spasmodic colic and ileal impaction colic in the horse. *Equine Veterinary J.* **30**, 194–199.

Putman, R.J., Pratt, R.M., Ekins, J.R. and Edwards, P.J. (1987) Food and feeding behaviour of cattle and ponies in the New Forest Hampshire. *J. Applied Ecology* **24**, 369–380.

Ralston, S.L. (1984) Controls of feeding in horses. *J. Animal Science* **59**, 5, 1354–1361.

Ralston, S.L. (1992) Regulation of feed intake in the horse in relation to gastrointestinal disease. *Europaische Konferenz uber die Ernahrung des Pferdes* **1**, 15–18.

Ralston, S.L. (1995) Postprandial hyperglycemica/hyperinsulinemia in young horses with osteochondritis dissecans lesions. *J. Animal Science* **73**, 184 (Abstract).

Ralston, S.L., Foster, D.L., Divers, T. and Hintz, H.F. (2001) Effect of dental correction on feed digestibility in horses. *Equine Veterinary J.* **33**, 390–393.

Raymond, S.L., Curtis, E.F. and Clarke, A.F. (1994) Comparative dust challenges faced by horses when fed alfalfa cubes or hay. *Equine Practice* **16**, 42–47.

Redbo, I., Redbo-Tortensson, P., Odberg, F.O., Hedendahl, A. and Holm, J. (1998) Factors affecting behavioural disturbances in raehorses. *Animal Science* **66**, 475–481.

Robinson, N.E. (2001) Report on the International workshop on Equine chronic airway disease. *Equine Veterinary J.* **33**, 5–19.

Rolgalski, M. (1970) Behaviour of horse at pasture. *Kon Plski* **5**, 26–27.

Ruckebusch, Y., Vigroux, P. and Candau, M. (1976) Analse du comportements alimentaire chez les equids. *C.R.J. d'Etude Cereopa, Paris*, 62–72.

Ruohoniemi, M.R., Kaikkonen, R., Raekallio, M. and Luukkanen, L. (2001) Abdominal radiography in monitoring the resolution of sand accumulations from the large colon of horses treated medically. *Equine Veterinary J.* **33**, 59–64.

Sandin, A., Skidell, J., Haggstrom, J. and Nilson, G. (2000) Postmortem findings of gastric ulcers in Swedish horses oder than age one year: a retrospective study of 3715 horses (1924–1996). *Equine Veterinary J.* **32, 1**, 36–42.

Savage, C.J., McCarthy, R.N. and Jeffcott, L.B. (1993) Effects of dietary energy and protein on induction of dyschondroplasia in foals. *Equine Veterinary J.* **S16**, 80–83.

Stanier, W.B., Akers, R.M., Williams, C.A., Kronfeld, D.S. and Harris, P. (2001) Plasma insulin-like growth factor-1 (IGF-1) in growing thoroughbred foals fed a fat and fiber versus a sugar and starch supplement. *Proceedings of 17th Equine Nutrition and Physiology Symposium*, pp. 176–177.

Stefansdottir, G.L., Hyslop, J.J. and Cuddeford, D. (1996) The in situ degradation of four concentrate feeds in the caecum of ponies. *Animal Science* **62**, 646 (abstract).

Stull, C.L., Hullinger, P.J. and Rodiek, A.V. (2001) Metabolic responses of fat supplementation to alfalfa diets in refeeding the starved horse. *Proceedings of 17th Equine Nutrition and Physiology Symposium*, pp. 159–160.

Tinker, M.K., White, N.A. Lessard, P., Thatcher, C.D., Pelzer, K.D., Davis, B. and Carmel, D.K. (1996) Assessment of risk associated with events in a prospective study of equine colic. *Proceedings of 42nd American Association of Equine Practitioners Convention* **42**, 332–333.

Tinker, M.K., White, N.A., Lessard, P., Thatcher, C.D., Pelzer, K.D., Davis, B. and Carmel, D.K. (1997) Prospective study of equine colic risk factors. *Equine Veterinary J.* **29, 6**, 454–458.

Topliff, D. (1997) Nutritional management of horses with Hyperkalaemic periodic paralysis. In *The Veterinarians Practical Reference to equine Nutrition*. Purina Mills and American Association of Equine Practitioners 167–171.

Townson, J., Dodd, V.A. and Brophy, P.O. (1995) A survey and assessment of racehorse stables in Ireland. *Irish Veterinary J.* **48**, 364–372.

Tyler, S.J. (1972) The behaviour and social organisation of New Forest Ponies. *Animal Behaviour Monograph* **5**, 85–196.

Uhlinger, C. (1990) Effects of three anthelmintic schedules on the incidence of colic in horses. *Equine Veterinary J.* **22**, 251–254.

Valentine, B. (1997) Nutrition and neuromuscular disease. In *The Veterinarians Practical Reference to equine Nutrition*. Purina Mills and American Association of Equine Practitioners, pp. 123–131.

Van Weeran, P.R. and Barneveld, A. (1999) Effect of exercise on the distribution and manifestation of osteochondritic lesions in the warmblood foal. *Equine Veterinary J.* **S31**, 16–25.

Warr, E.M. and Petch, J.L. (1992) Effects of soaking hay on its nutritional quality. *Equine Veterinary Education* **5**, 169–171.

Weise, J. and Lieb, S. (2001) The effects of protein and energy deficiencies on voluntary sand intake and behaviour in the horse. *Proceedings of 17th Equine Nutrition and Physiology Symposium*, p. 103.

Willard, J.G., Williard, J.C., Wolfram, S.A. and Baker, J.P. (1977) Effect of diet on cecal pH and feeding behaviour of horses. *J. Animal Science* **45**, 87–93.

Williams, C.A., Kronfeld, D.S., Stanier, W.B. and Harris, P. (2001) Glucose and insulin responses in thoroughbred mares are influenced by reproductive stage and diet. *Proceedings of 17th Equine Nutrition and Physiology Symposium*, pp. 178–179.

Chapter 4

HOUSING, MANAGEMENT AND WELFARE

D.S. MILLS
Animal Behaviour, Cognition and Welfare Group, University of Lincoln, Lincolnshire School of Agriculture, Caythorpe Campus, Lincolnshire NG32 3EP, UK

A. CLARKE
Faculty of Veterinary Science, University of Melbourne, Australia

Abstract. Horses tend to be housed in loose boxes, stalls, barns and shelters for ease of management, however these systems present several possible threats to equine health and welfare. These systems are reviewed together v ᶜʰ; the concerns they raise. A common system for the evaluation of the welfare of contained animals focuses on the provision of five freedoms. These are freedom from hunger, thirst and malnutrition, from discomfort, from pain, injury and disease, from fear and distress and to express most normal patterns of behaviour. This approach is used to assess the ways in which horse welfare may be compromised by certain housing practices and management regimes. Recommendations as to how these problems can be resolved and to promote good practice are provided.

1. Introduction

"It does not require any vast expenditure of thought to discover that life is action, 'to be' is synonymous with 'to do': therefore, it is a sheer necessity of existence that an animated being must be doing something. Such is the primary consequence of existence. Thus to breathe and to move imply one act; since, if the lungs cease to dilate, respiration immediately terminates, and, with it animation comes to an end. Yet it remained for mortal perversity to rebuke the first principle of established philosophy, when stables were built, in which a breathing animal was to be treated as it were an inanimate chattel."

These strong words were written by James Lupton over 100 years ago (Lupton 1884) but are still a powerful reminder of the two main problems facing the stabled horse today: the aerial environment and the psychological impact of the stable on the horse. These factors therefore form the main focus of this chapter.

There are unique issues to be dealt with and challenges faced in housing horses, especially in comparison with other species. It is possible to extrapolate some principles of housing from farm animal species, however, there are situations where this approach is detrimental. For example, maintaining a warm environment for pigs and poultry can increase their production efficiency. There is a respiratory cost to the animals with this approach. However, even minor levels of respiratory disease can influence the performance of an equine athlete. In this situation respiratory health is more critical for horses than meat production animals. Horses are also generally longer living compared with the majority of farm animals, which are slaughtered when relatively young. Conditions such as the allergic respiratory

N. Waran (ed.), The Welfare of Horses, 77–97.

disease, chronic obstructive pulmonary disease (COPD or heaves) are primarily seen in older horses, ranging from best loved children's ponies to the most valuable bloodstock. Paying attention to the quality of the air which horses breath has both short-term and longer term health and welfare implications for horses (Clarke 1987; Holocombe *et al.*, 2001).

Horses are also unique in the relatively low stocking density at which they are housed. Stocking density of intensively housed farm livestock has a major effect on the quality of the air that they breathe. By comparison, the quality of air horses breathe is primarily affected by the quality of the food and bedding that is provided and the management of bedding.

Shelter, grazing, shade and a source of water, are believed to be the key resources, that determine the core area requirements in the free-roaming horse (Tyler 1972). However, modern housing tends to focus on the provision of a safe, clean and cost-effective environment in which an animal can be managed conveniently and in a cost-effective way. Neither these owner-centred considerations nor the commonly cited justification for the need to conserve grazing, really recognise the health and welfare implications of housing a horse, which have the potential to be quite serious. Indeed horses may prefer not to be housed even in inclement weather. Schatzmann (1998) found that 5 horses provided with free access to a 15 acre paddock and box stall with straw bedding, hay and water all chose to stay outside during the Swiss winter, as long as some grass was available.

The evolutionary history of the horse reflects selection for a social species living in open plains where flight is the primary mechanism of escape from predation. It should not be surprising therefore, if there were to be an inherent aversion by the horse to the isolation and confinement associated with many housing systems (Mills & Nankervis 1999). Housing also poses a biological challenge to the normal mechanisms of health regulation in the horse. Enclosure inevitably involves an accumulation of potentially harmful substances in the aerial environment, by virtue of the restriction it imposes on the circulation of air around an active biological system (*i.e.* the horse, and its immediate environment). Recent studies (Jackson *et al.*, 2000; Holcombe *et al.*, 2001), have demonstrated that even in apparently healthy animals a greater level of upper and lower airway inflammation is seen when housed, and that this subsides when the animals are turned out. This suggests that housing may both psychologically and physically stress the horse, which, if not properly managed, can rapidly result in welfare problems. Different systems and practices pose different threats and so we start with a brief review of the systems available before we consider their impact on the psychological and physiological well-being of the horse.

2. Housing Systems and Their Impact on the Horse

Housing may be divided into four broad categories: stalls, loose-boxes, barns and shelter systems (Clarke 1994). Brief details of each, together with its effect on the behaviour of the horse are given below.

2.1. LOOSE BOXES (BOXSTALLS)

Loose-boxes are probably the most common form of housing used in Europe and North America, since they allow each animal to have its own space and personal management routine according to the wishes of individual owners. Boxes usually vary in size from 3 m by 3 m approximately for those designed to hold a pony, to 3.6 m by 3.6 m for larger horses and 5 m by 5 m for a foaling box. This allows the horse to walk around in the box, but provides a degree of confinement, which may restrict the normal level of movement of the occupant (see Figure 1). Social isolation may also be a problem in loose boxes which are often separated by solid full height walls. This sort of partition is commonly justified by the need to prevent cross infection of airborne pathogens. Alternatively, walls may contain grills or only be partial, to allow greater circulation of air and some degree of social interaction between boxes. Loose boxes may be part of a larger enclosed building with rows of units separated by a central passage. This type of building is colloquially known as an American Barn. In this case they all share a common airspace. Whilst this is commonly thought to be beneficial to the horse, through the provision of a less variable temperature and dry conditions, such buildings can be very difficult to ventilate effectively. The biggest challenge is ensuring that there is effective distribution of air throughout the building and that all stalls are adequately ventilated (Clarke 1987). Kiley-Worthington (1987) concluded that horses housed in loose-boxes even on ad libitum rations, spent more time standing (40% of time recorded)

Figure 1. The stable/loose-box can restrict a horse's normal behaviour (courtesy of Natalie Waran).

and less time feeding (47% of recorded time) compared to free roaming horses. Feral horses have been observed to spend on average 20% of their time standing and 60% of their time eating (Duncan 1980), whilst domestic horses with free access to pasture are reported to spend more than two thirds of their time feeding (Crowell-Davis et al., 1985). Boxing therefore interferes with the normal allocation of diurnal activity. Rees (1984) has suggested that the tendency to arrange boxes so that they face the main focus of activity interferes with the ability of horses to relax, and this may, at least in part, explain this effect. Thus on-going maintenance behaviours, such as feeding, may be continually interrupted by vigilance behaviour. Further, the arrangement of boxes around a central courtyard where they can see but not approach other horses, whilst often thought to be of benefit to the horse, may actually increase behavioural frustration or the aversiveness of the environment for some individuals (McAfee et al., in press).

2.2. STALLS

Stalls and tethers involve tying animals via a head collar under some form of cover. Horses should be tied so that they can still lie down, but this system still imposes much greater restriction on movement. Partitioning may be such that animals are individually isolated, paired or in larger bays, but the dividing walls are not usually solid to the full height of the building. Therefore social isolation may not be as great as in many loose boxes, since horses are normally in closer proximity and able to view more of neighbouring conspecifics. Incidents of aggressive behaviour between horses in adjacent stalls can be managed by moving one or both of the individuals involved. A typical stall may be 2–2.5 m wide and 3 m long and should slope towards a drain and passageway at the rear of the horse. The slope should be the minimum required for effective drainage in order to reduce strain on the flexor tendons. The advantages of this system include the reduction in the space and bedding requirements and the resultant speed with which animals can be mucked out. However, the level of confinement imposed by this system, and the inevitable need to approach horses from behind are common causes for concern. Accordingly the system tends to be used where space and/or time is at a premium and with horses that are out at work for a large part of the day. Interestingly, horses kept in this system with ad libitum forage are reported to spend approximately 65% of their time feeding and 25% standing (cited in Ogilvie-Graham 1994), which is more similar to the reported time budgets of free roaming than loose housed horses. However, as expected, less time is spent lying in this system (3–6%, Ogilvie-Graham 1994) than in others (generally reported to be around 10%).

2.3. BARNS/COVERED YARDS

There are two basic types of barns. The first is the so-called 'American-barn' where horses are housed in individual loose-boxes under one large roof (see above). The second system is the horse-barn in which horses are group housed. This system is cost effective and also relatively low maintenance. The horses may not be so clean and some aspects of their health are more closely dependent upon that of other members of the group. The common airspace, high stocking density and tendency to use deep litter may all increase the risk of spread of disease, should it enter the system. Since horses are free to move and interact, it is not surprising that yarded horses fed ad libitum forage, have similar time-budgets to their feral counterparts (Kiley-Worthington 1987). The main concern with this system relates to the risk of injury from aggression, since stocking density may not allow an individual to withdraw from an agonistic encounter. This sort of interaction is more likely in an unstable social group, such as that which may exist on a public livery yard where there may be a frequent turnover of stock. However, large personal yards, foaling centres and horse farms can group together horses of similar age or individuals well known to each other in order to minimise this risk.

2.4. SHELTERS

Finally, there is the shelter and paddock system. This comprises of a simple, largely open-fronted building, which opens onto a paddock. It may vary from an extensive grassy paddock designed for several horses (field management system) to a more restrictive individual sand paddock (dry lotting). In the latter system, a yard may contain several units in a block for ease of management and there is much greater control over factors like feeding and cleanliness. All of these types of system are potentially cost effective and relatively low maintenance. Social isolation is obviously potentially less and the individual dry lotting system may circumvent many of the problems associated with group housing, although horses may still bite each other across a fence. Movement restriction is also less in these systems, but grass surfaces may be prone to poaching and spoiling in cold, wet conditions.

Further details on the principles of construction and design can be found in Sainsbury (1987).

3. Assessing the welfare implications of current equine management and housing practices

3.1. WELFARE, HEALTH AND SUFFERING

Welfare is not a simple physical characteristic that can be easily quantified, but efforts can be made to assess it as objectively as possible. Historically, concern for the welfare of captive animals has focused on meeting the physical requirements of the organism, such as the need to provide a balanced diet and prevent physical illness (Harrison 1964). However, if it is accepted that animals can suffer when such needs are not met, (*i.e.* at times of physical stress) then it is accepted that animals are not automatons but have some degree of awareness. If this awareness extends to an awareness of the environment, it follows that animals may suffer as a result of environmental stressors, many of which have no physical effect on the animal (psychological stress). There are many ways in which the management of horses might raise such concerns. Stabled horses are often deprived of the opportunity to interact fully with the environment (*e.g.* through confinement and isolation) and they are often unable to apply the mechanisms, which have evolved for the regulation of normal ongoing activity (see Figure 1). An example of this is where the provision of the animal's maintenance requirements is by unnatural means, such as the provision of nutrients in the diet, in a concentrated form.

3.2. THE FIVE FREEDOMS

One way of assessing the diverse threats to an individual's well-being is embodied in the Farm Animal Welfare Council's 'Five Freedoms' (FAWC 1992). This suggests that all domestic animals should have:

- Freedom from hunger, thirst and malnutrition;
- Freedom from physical and thermal discomfort;
- Freedom from pain, injury and disease;
- Freedom from fear and distress;
- Freedom to express most normal patterns of behaviour.

Different forms of housing and management impact on each of these measures of welfare, and optimal practice is further obscured when we recognise that there are specific psychological dimensions to each measure. Scientific investigation helps to identify genuine areas of potential concern, rather than the popular anthropocentric welfare priorities. This point is illustrated further with respect to each 'freedom' below.

3.2.1. *Freedom from hunger, thirst and malnutrition*
The provision of a balanced and nutritious diet together with ad libitum water may be essential but is not sufficient to meet this freedom. Meal feeding (as experienced by intensively managed horses) is not the evolved method of ingestive behaviour and so its imposition may pose problems for the domestic horse (Mills 1999). Ralston *et al.* (1979) noted that horses do not voluntarily starve themselves for more than 4 hours and Krzak *et al.* (1991) found that wood chewing tended to occur predominantly at night when horses were likely to have empty stomachs. It seems likely therefore that the evolved mechanisms designed to regulate intake, may not be appropriate to restrictive feeding practices. Ad libitum feeding rather than more frequent meals would appear to be preferable since McGreevy *et al.* (1995) found that the provision of forage more than three times a day was associated with a greater risk of problem behaviours than less frequent feeding.

Less dominant individuals in group housing management systems may also be at a disadvantage to more dominant individuals in terms of gaining access to feed and water. Multiple feed and water troughs will help to avoid this problem.

Malnutrition not only refers to an inadequate supply of nutrients but also the provision of nutrients in a form that may cause harm. Concentrate feeding may meet an animal's chemical nutrient requirements, but the use of concentrate to increase the energy density of ingested food may not be good for a horse's well-being. Concentrates are associated with an increase in gastro-intestinal acidity (Rowe *et al.*, 1994, Johnson *et al.*, 1998) which may predispose animals to gastric ulceration (Murray *et al.*, 1996; Pagan 1997), wood chewing and crib-biting (Johnson *et al.*, 1998).

The provision of ad libitum water may also not guarantee freedom from thirst, since Welford *et al.* (1999) found that a shift in the timing of the normal management schedule of Thoroughbreds resulted in a drop in daily water intake of over 7% on average. Owners should be aware that changes in management may result in disturbances to behaviour which might be of concern.

The subject of nutrition and welfare is discussed further in Chapter 3.

3.2.2. *Freedom from physical and thermal discomfort*
Complex problems arise with the provision of physical and thermal comfort, since horses have been shown to exert bedding preferences in experimental situations which may be a risk to their physical health (Mills *et al.*, 2000). Mills *et al.* (2000) found that straw bedding was preferred over wood shavings and both of these were preferred to shredded paper. Whilst straw bedding is associated with fewer behaviours of welfare concern (McGreevy *et al.*, 1995), it is also associated with an increased risk of respiratory disease.

Horses do not require warm housing and maintaining adequate ventilation is almost certainly more important than maintaining a higher ambient temperature. With acclimatisation, adult horses have been shown to be able to comfortably tolerate temperatures as low as $-10\ °C$ (McBride *et al.*, 1983) and even 2-day old foals may tolerate temperatures as low as 5 °C as long as they are well fed (Clarke 1987). Quartz halogen radiant heaters offer a practical source of extra heat where it is required for foals.

Good ventilation should keep the housing environment free of damp and should not generate drafts. Concern over the risk of chilling is sometimes expressed for horses, which have had their coats clipped for winter. This is done to reduce sweating up during physical work and consequently the risk of chilling when this stops, in addition to making cleaning less laborious at a time when the horse's coat may be more prone to soiling. For these individuals it is preferable to use rugs and extra layers of blankets, rather than reduce ventilation of the stable to maintain thermal comfort. Thermally efficient materials are also increasingly being used for horse rugs (Clarke 1994).

3.2.3. *Freedom from pain, injury and disease*
This implies the need for safe, secure well-constructed housing, without dangerous fittings etc. However, as already mentioned, bedding substrates preferred by the horse may not be preferred by the owner, because of the risk of respiratory disease or the need to manage such a problem.

Individual housing may be preferable in terms of reducing the risk of injury between horses in an unstable group, but this should be balanced against any need for social contact. It is often thought that more restrictive housing such as tie stalls may predispose horses to tendon injuries, but the authors are not aware of any scientific data to support this supposition.

3.2.4. *Freedom from fear and distress*

The barrier associated with enclosure and captivity may result in some problems with this freedom for an animal adapted to living in expansive open areas, with flight as the main mechanism for defence. Where animals are group housed the barrier may not allow an individual to withdraw to a safe distance from a conflict situation and result in individuals which live in fear of aggression. In individual housing systems distress may result from the greater level of confinement and social frustration. Chronic frustration may cause a psychological reaction in the form of stereotypic behaviour (see Chapter 5) and possibly increase aggression as well. Stereotypic behaviours are repetitive, relatively invariate behaviours with no obvious function from the context in which they are performed (Mason 1991). They include behaviours such as weaving, box walking, cribbing and windsucking in the horse, which are relatively common in captivity with reported prevalences of up to 8.3% for crib-biting, 9.5% for weaving and 7.3% for box-walking (Nicol 1999). They have not been reported in animals which have always been free-roaming. Several housing factors have been associated with an increased risk of stereotypy including: reduced social contact, systems with less than 75 horses and the absence of a paddock (McGreevy *et al.*, 1995). Luescher *et al.* (1998) also found that weaving is more common in yards with a smaller proportion of standing stalls.

Many owners find these behaviours aesthetically unacceptable and try to prevent them from being performed if they occur in an individual (McBride & Long 2001). However, it is likely that these behaviours are highly motivated and an expression or consequence of distress rather than a cause of distress to the horse. Therefore their physical prevention, such as through the elimination of cribbing surfaces and management aids such as cribbing collars and antiweaving bars may increase suffering (McGreevy & Nicol 1998; McBride & Cuddeford 2001).

3.2.5. *Freedom to express most normal patterns of behaviour*

Providing freedom to express most normal patterns of behaviour can be particularly problematic, not least because the term normal has several meanings, none of which necessarily relate directly to welfare (Cooper & Mills 1997). Behaviour may be considered normal in a natural sense, because it is found in the normal wild population, in a statistical sense because it is relatively common among domestic horses; normal in a functional sense because it is adaptive or optimal; and normal in a social sense because it is culturally acceptable. There are clearly some behaviours which the wild horse expresses which are not desirable for good welfare in the domestic animal such as those associated with predator avoidance; and so 'normal pattern' should not be equated with 'natural range'. Rather, it is important that the naturally evolved mechanisms for regulating the natural range of behaviours are not put under strain by the domestic situation, due to environmental deprivation and frustration. It is therefore essential that the organisation and motivation of behaviour is understood. As we have already seen there is more than the physical requirements for survival to be met for good welfare.

The term 'normal pattern' may also be used to reflect the structure of a specific functional behaviour, *i.e.* the way in which the goal of a behaviour is achieved.

Deviation from the normal evolved state may also (but not necessarily always) result in welfare problems. For example, feeding forage may be a substitute for some of the problems associated with feeding concentrate discussed above, but the way it is fed (*i.e.* from a hay net) may have welfare implications which can be overlooked quite easily. Feeding from a hay net reduces wastage and eases management, but the posture is abnormal, and the additional time spent with the head raised may compromise the function of the mucociliary escalator of the upper respiratory tract (Racklyeft & Love 1990). This structure is an important part of the body's defence against disease, trapping potential disease causing particles before they can cause a problem. Thus feeding from a hay net may also predispose an individual to respiratory infection.

The time that a behaviour occupies may also be considered a part of its 'normal pattern' and may be cause for concern, as we have already seen. By way of further example, not only may housing type, but also bedding substrate may affect the time spent lying down or asleep. Outside certain functional limits, this may predispose an individual to both physical and psychological illness (Mills *et al.*, 2000). Further, the feeding of concentrates allows a horse to achieve its nutrient intake in an abnormally short time compared to the wild state. It has been estimated that a wild horse would normally chew around 57,600 times in a day (Cuddeford 1999) and the production of saliva is linked to the process of chewing, rather than presence of food related stimuli (Alexander & Hickson 1970). Saliva not only helps to lubricate the food and aid swallowing, but is also alkaline and may therefore be an important buffer against increases in gastro-intestinal acidity, such as that which occurs with concentrate feeding (Nicol 1999) or more generally help regulate gastro-intestinal pH. Therefore not only may the chemical composition of concentrates affect gut pH directly, but also the reduction which they cause in time spent chewing may deprive the animal of one of its main mechanisms for regulating gut acidity during the digestive and absorption processes. Ulceration may be one of the obvious physical welfare problems which arise as a result, whilst cribbing and woodchewing may be behavioural expressions of an attempt to compensate. This is discussed further in Chapters 3 and 5.

In summary, if the horse is expected to behave in an unnatural way, it is important to pause and consider what the implications of this may be for the horse's welfare, rather than simply assume that because the horse still functions it can adapt without compromising its well-being.

Further scientific investigation is undoubtedly necessary to quantify these problems further, but in the final analysis, an ethical judgement must be made about what is an acceptable level of risk or compromise, if horses are to be kept in captivity. In the next two sections we consider the nature of the major risks to physical and psychological health posed by captive housing and management.

4. Reducing Risk to Physical Health

Respiratory health is the major physical health concern for the stabled horse. Respiratory diseases can be categorised as being either acute or chronic and can occur in degrees ranging from life threatening, *e.g. Rhodococcus equi pneumonia* in foals, to very mild degrees of covert small airways disease, which is only manifest as a loss of performance when the horse is under extreme exertion. These mild degrees of airway disease in early life may be the precursor of the more debilitating disease COPD in later life. Special attention must also be given to design features which decrease the risk of physical injuries (Clarke 1994). One condition that should be highlighted for breeding farms is sesamoid fractures in young foals. This is the major preventable fracture that foals suffer and can have consequences for the foal's future life. This occurs when young or immature foals run to exhaustion. The incidence of this problem can be decreased by ensuring that mares and foals at risk are kept in smaller paddocks until the foal strengthens up. It is also important to avoid placing younger, weaker foals and their mothers in groups with older stronger foals which are capable of more exercise.

4.1. INFECTIOUS AND PARASITIC RESPIRATORY DISEASE

Horses are prone to a wide range of respiratory tract infections including viruses, bacteria and mycoplasmas. A review of these agents is beyond the scope of this chapter (Cullinane 1997). Careful monitoring of horses for early signs of infections, such as daily taking of temperatures and the early instigation of veterinary treatment, is critical in lessening the impact of infections. Rigid adherence to a vaccination programme is also critical. While isolation and quarantine practices can be of benefit in some situations, *e.g.* the separation of young stock and brood mares, this approach is not always practical or productive, especially in stables where horses are regularly travelling to events and competitions.

Environmental control alone is unlikely to limit the spread of a highly contagious respiratory disease in a susceptible population of horses. This is especially the case where the horses are the primary sources of the pathogens and where these infectious agents do not survive for long periods of time away from the horse. This occurs for example with an outbreak of influenza virus or herpes virus infection (Clarke 1987). This situation is further complicated with diseases such as Equine Herpes Virus infections and Strangles where there can be asymptomatic carrier horses which intermittently shed the infectious agents. However, while the air quality of the stables may not affect the incidence of these diseases it can affect both the duration and the severity of disease in individual horses. Paying attention to the air quality of the stables will result in horses getting better quicker and decrease the loss of training and competition days.

Environmental control of infectious and parasitic respiratory disease can be very successful where the agents survive and sometimes even proliferate outside of the host. *Rhodococcus equi* pneumonia is a clear example (Wilson 1997). *Rhodococcus equi* is an actinomyecte which proliferates in equine faeces in warm weather and becomes airborne to be inhaled by foals in dry dusty conditions in both paddocks and to a lesser extent in barns. Removal of faeces from paddocks and barns and damping down dusty areas in paddocks or avoiding such areas with foals will decrease the incidence of this disease in endemic areas.

Lung worm (*Dictyocaulus arnfieldi*) is well recognised for causing respiratory disease in horses. Migrating larvae of other parasites including *Parascaris equorum* and *Strongyloides westeri* can also cause respiratory disease especially in young horses. The control of parasites in horses requires careful pasture management and the regular use (and notation) of appropriate anthelmintics (Bailey 1992). The eggs and larval stages of most equine parasites are long-lived and resistant to desiccation. In relation to stable management, droppings should be regularly and thoroughly removed and deep litter management avoided.

4.2. INFLAMMATORY AIRWAY DISEASE

It is the horse's small airways, the bronchioles, which are most affected by the stable environment (Raymond & Clarke 1997). COPD is the best known manifestation of small airway disease in horses. Symptoms of COPD include chronic cough, flared nostrils, forced abdominal breathing and severe exercise intolerance. Inflammation, increased mucus production and bronchospasm brought on by a full-blown allergic reaction to inhaled mould spores cause these clinical signs. At the other end of the scale there is covert lower airway disease known as Lower Respiratory Tract Inflammation (LRTI) or Small Airway Disease (SAD). This is also an inflammatory process but not believed to be a full-blown allergic reaction as seen in COPD. There may not be overt manifestation of LRTI other than a loss of performance in equine athletes. Diagnosis of LRTI often requires veterinary endoscopic examination and an analysis of mucus samples collected directly from the horse's airways.

In this section emphasis will be placed on providing practical management approaches and building design criteria to help prevent and alleviate overt and covert manifestations of small airway disease.

The successful environmental control of respiratory disease necessitates that the horse's exposure to the pathogens be kept below the Threshold Limiting Value (TLV) which will induce disease (Clarke 1993). Unfortunately the TLV for horses to stable dust such as mould spores or noxious gases such as ammonia is unknown. Furthermore the TLV can vary both within and between individuals. For example the TLV for inhaled dust will be greatly decreased in a horse which is suffering from a viral respiratory tract infection and has damaged cilia in its airways. The cilia work to clear inhaled dust from the lungs and when the cilia are damaged there can be the equivalent of accumulation of inhaled particles for the lungs to deal with. This increase of particles retained in the lung as a result of infection is believed to explain, in part at least, why animals (including humans) are prone to develop lung allergies after a bout of infectious disease.

The best management approach in relation to LRTI is to minimise the horse's exposure to airborne contaminants at all times. This necessitates that both the sources of airborne contaminants including the feed and bedding and the processes regulating their removal, primarily ventilation, be focused on. It is also critical to focus on the horse's breathing zone and the challenge posed at this level. For example a particles-counter or ammonia detection kit held five feet above a straw bedding in a well-ventilated stable may give a low-test result. However, high levels of dust and ammonia could be inhaled when the horse nuzzles into its bedding. This point has been reflected in recent years in research with a move away from static air samplers to units, which can be clipped onto the horse's head collar. These latter units provide data, which is collected from the horse's 'breathing zone'.

4.3. FEED

Equine nutrition is covered in more detail in Chapter 3, however it is worth noting that many of the welfare problems that are associated with the housing of horses, are related to the feeding types and methods that are used. Hay is the most common source of respirable dust and mould spores that horses are exposed to. The primary factor, which affects the mould spore content of hay (and straw), is the moisture content at baling. Hay and straw that are baled with a high moisture content undergo heating in the first month after baling. During this process there are several species of fungi and actinomycetes, which develop. These species produce large numbers of spores less than five microns in diameter, which are capable of being inhaled to the deepest levels of the lungs. Eventually horses develop an allergic response to these spores which is manifest as COPD. However, these spores can also induce non-allergic inflammation of the airways in young horses. This is a separate condition from COPD. COPD is a rare condition in temperate regions such as Australia and parts of the USA. One of the main reasons for this is that hay and straw are more likely to be baled at a lower moisture content in Australia and there is less mould contamination in these source materials. Laboratory based tests are available to assess the moulding of hay and straw. This is important because the eye and the nose are not good judges of the levels of mould contamination (Raymond et al., 1997).

The soaking of hay is a well established method of minimising a horses exposure to mould spores. It is essential that the hay is wet throughout as small dry pockets can still lead to significant challenges of dust. There is also some evidence that soaking of hay for more than half an hour can lead to the leeching of water soluble vitamins and other nutrients. Also while soaking of hay will decrease the inhalation of spores it is possible that significant levels of mycotoxins could be ingested by the horse (Raymond et al., 2000). Mycotoxins are metabolites produced by moulds and can have a wide range of effects ranging from decreased reproductive efficiency to nervous system damage. Soaking of hay is also labour intensive and is not practical in colder climates such as a Canadian winter.

There is a range of products, which are alternative sources of forage for horses. These include specially produced mechanically dried hay, chopped moistened complete diets, silage type products and hay cubes. Research has shown that horses eating all of these products inhale less dust than when eating traditionally produced hay.

One way of producing hay which does not ferment (or 'heat') is to mechanically dry the hay to a moisture content of less than 10% before it is baled. There are two approaches to this process. In both situations the grass is generally cut and allowed to dry in the field for between 12 to 24 hours. Following this initial drying period it is picked up loose in the field and taken to either a high temperature dryer where it is dried and baled in the one process or alternatively taken to large barns and loosely stacked with air blown through the grass to dry it before it is baled. Both of these approaches produce a premium product, which is more costly than traditionally produced hay because of equipment costs, the extra handling and energy costs. Acid based additives have also been promoted to prevent the moulding of hay (Clarke 1994). These are applied as the hay is baled. However, the success of the agents can be limited and they can also affect palatability of hay.

Haylage and large bale silage have widespread usage in the United Kingdom and their use is spreading into North America (Raymond et al., 1997). Haylage and silage are baled with a higher moisture content and then sealed in airtight plastic bags. A fermentation response occurs just after baling and the associated increase in acidity inhibits the development of moulds and bacteria. The primary distinction between haylage and silage is the maturity of the grass or legume when it is cut.

Young grasses tend to be conserved as big bale silage and some horses can have loose dropping when fed this product. Some owners do not find this acceptable. Haylage tends to be prepared in small bales and more mature grasses are used. The high sugar contents of mature ryegrass makes this a popular choice for haylage, especially in the United Kingdom.

Once a bale of silage or haylage is opened it should be fed within a few days. This is because moulding will develop especially in warm weather. There have been rare incidences of botulism associated with the feeding of large bale silage to horses. Any bales of silage which are punctured or which have signs of being contaminated with dirt or the remains of animals e.g. dead rabbits, should not be fed to horses.

Hay cubes are very popular alternatives to hay especially in North America (Raymond et al., 1994). There are two basic forms of cubes. The first is produced from previously baled hay. The second is made from grass which is freshly cut and dried in the paddock for between 12 to 24 hours before being mechanically picked up and taken for drying and cubing in one process. The quality of hay cubes used in the former process is dependent on the quality of hay used. The latter process is beneficial in that there is a decreased risk of mycotoxin for horses. Lucerne is the most common commodity used for hay cubes, however, there are lower energy varieties with grasses as well as higher energy products which incorporate maize with lucerne. The latter is popular for performance horses.

All of the forage alternatives described above expose horses to lower dust and respirable mould spores than dry hay. There can be significant differences in the cost-effectiveness of these products based on a dry-matter basis. From a behavioural point of view hay cubes and haylage products can be eaten a lot quicker than hay, which can lead to problems with stereotypical behaviour. This can be avoided by forcing the horse to take extra time eating its forage. One way of achieving this is feeding haylage in a hay net with small holes.

4.4. BEDDING

Bedding is the second major source of respirable mould spores in the stable. The level of mould contamination varies greatly between different batches of straw. As with hay the moisture content at baling is the key factor which influences the level of moulding in the straw and the level of exposure the horses face. However, even the cleanest of straw contains significantly more respiratable spores than alternatives such as wood shavings, paper, sawdust, peat or synthetic products. Significant levels of mould contamination can occur in wood shaving, which are produced with high moisture levels or exposed to the elements. A common sign of moulding in wood shavings in plastic bags can be grey streaks through the products.

Bedding horses on alternatives to straw does not ensure that the bedding is not a significant source of mould spores or other pathogens. Very high levels of airborne respirable spores can emanate from bedding materials, which are managed in a deep litter approach, and also in stables which are poorly ventilated. While different species of moulds tend to develop on wood shavings compared with straw, these species still produce large numbers of tiny spores, which can reach the horse's small airways.

At the end of the day using alternatives to straw such as paper or wood shavings will improve the respiratory health of horses. However, care must be taken to ensure that moulds do not grow in sites and that ammonia levels do not build up.

4.5. AMMONIA

Ammonia is the most common noxious gas to which horses are exposed (Curtis *et al.*, 1996). It is released by the action of bacteria on horse's urine and faeces on the floor of the stable. Ammonia damages the mucociliary escalator and increases mucus production. Feeding of high levels of protein can increase ammonia production in a stable but there is considerable individual variation in this context. Ammonia levels increase in poorly ventilated barns especially where drainage is poor and deep litter management is practical. Ammonia levels also generally increase with increasing temperatures and humidities. Methods, which decrease ammonia levels in stables, include:

- Improving drainage;
- Ensuring adequate ventilation;
- Frequent removal of excreta and wet bedding;
- The use of commercially available ammonia control products.

4.6. VENTILATION

A well ventilated stable will help to decrease the horse's exposure to a wide range of pathogens including; noxious gases, dusts and microbes (Clarke 1994).

Ventilation can also be used to aid in decreasing the risk of moulding of plant-based bedding materials. However, ventilation cannot be used to overcome inherent management problems. For example, a horse can still inhale significant levels of spores from mouldy hays or bedding in a well ventilated stable.

4.6.1. *Natural forces of ventilation*
The majority of horse stables and barns can be ventilated effectively without the use of mechanical support.

There are three natural forces of ventilation:

i) The stack effect. This occurs as a result of warm air rising. The heat generated from horses can be harnessed to capitalise on this effect.
ii) Aspiration. Wind blowing across a roof of a building will suck air out of the building.
iii) Perflation. Air movement associated with wind blowing from end to end or side to side of a building. In exposed locations this natural force of ventilation can lead to drafts and must be compensated for with the strategic location of window, vents and draft dampers.

The ventilation of a building is most tested in still air conditions when the only driving force is warm air rising off the horses. Thus, stables should be designed on the assumption that windless conditions prevail. The relationship between levels of airborne contaminants and ventilation is curvilinear with the concentration of airborne contaminants being directly related to the reciprocal of the ventilation (see Clarke 1987). Thus the concentration of airborne contaminants increases sharply at low ventilation. Equally, doubling the air change rate at higher levels of ventilation is not associated with a halving of the airborne contaminants. In considering the principles of natural ventilation in still air conditions a target of four air charges per hour with the top door of the loose box or the main doors of a barn closed should ensure adequate ventilation all year round.

The principles that govern the natural ventilation of livestock housing have been described by Bruce (1978) and reviewed by Clarke (1987). The key elements to consider include:

• The stocking density and the heat production
• Insulation
• The height between the inlet for fresh air and the outlet for warm air
• The size of the inlets and outlets

The following guidelines (see Table 1) are provided for 'typical' loose boxes or barns based on the requirement for 4 air changes per hour.

The main advantage of having an insulated barn is that it decreases the size of the openings necessary to effectively ventilate the barn. Insulating the building will also decrease the risk of condensation.

Consideration must be given to the distribution of the inlets and outlets in stables. For individual loose boxes with a monopitched roof there should be one opening in the front and one in the back. There should be an additional outlet in the form

Table 1. Requirements for adequate ventilation of insulated and uninsulated horse housing.

	Insulated		Uninsulated	
	Loose box	Barn	Loose box	Barn
Dimensions per horse (m³)	50	85	50	85
Height between inlet and outlet (m)	1	1	1	1
Required inlet area/horse (m²)	0.27	0.38	0.34	0.46
Required outlet area/horse (m²)	0.14	0.19	0.17	0.23

of a chimney or covered ridge for stalls or barns with peaked roofs. Baffling can be used to decrease draughts or the entrance of rain or snow into vents on exposed walls. A more widespread distribution of vents is required for large barns. Large areas of Yorkshire boarding, capped ridges and 'breathing roofs' can also be very valuable in ensuring adequate ventilation and mixing of air in barns.

5. Reducing Risk to Psychological Health

The welfare of a subject relates to its subjective experience given the range of challenges it faces at any given time. A number of cross-sectional epidemiological studies have identified risk factors associated with threats to psychological well-being which may result in stereotypic behaviour (see Nicol 1999 for a review). These studies identify associations but cannot identify causal links. Thus the finding that horses that weave are more commonly associated with housing that allows minimal social contact does not prove that lack of social contact is a cause of this problem. Although there is no scientific support for the supposition that weaving can be copied by horses, many owners feel that this is the case (McBride & Long 2001) and so horses with this problem may be placed in more isolated positions because of their behaviour. This would result in the same finding. Other housing and management factors which have been associated with an increased risk of welfare problems have been mentioned earlier (see also Chapter 5). So in this section we will focus on the level of isolation and on feeding practice. Whilst confinement (restriction of movement) is commonly thought to be a major concern, the scientific evidence for this is still weak; this will be discussed in the latter part of this section.

5.1. ISOLATION

Isolation is a restriction on interaction with the environment. The issue of whether or not animals have a need to express certain behaviours in the absence of any stimulus for the behaviour remains controversial, but there is evidence to suggest that social isolation is stressful for the horse (Mal *et al.*, 1991; Jezierski 1992). This has adaptive value since social tendencies are an advantageous species-specific trait for the horse (Mills & Nankervis 1999). Selection would then favour any mechanism, which motivates effort to establish social contact when isolated. In this case, we hypothesise that the horse housed in social isolation is in a chronic state of frustration, which might be alleviated by social contact. Interestingly, Cooper *et al.* (2000) found that weaving in horses was significantly reduced when they had access to a conspecific in an adjacent stable through a grilled 1 m^2 portal. This effect has been replicated with the use of a similar sized mirror (Mills & Davenport 2002) and in a longer term study, McAfee *et al.* (in press) found that this sort of mirror also reduced aggressive threatening behaviour over the stable door. This change is also consistent with the hypothesis since chronic frustration from a signalled reward is often expressed in terms of increased aggression (Rolls 1990). There is therefore a strong case for the provision of social or 'pseudo-social' housing. The one situation in which isolation can be beneficial is for the foaling mare, since even in the natural state the mare is separates from the rest of the herd at this time (Tyler 1972). Isolation at this time may also reduce the risk of problems with the development of the sequence of motor patterns that are required for successful suckling by the foal (Mills & Nankervis 1999).

5.2. FEEDING PRACTICE

This describes what is fed and how it is delivered. Some of these issues have already been discussed earlier in this chapter, especially the problems associated with the feeding of concentrate. Recent unpublished studies on nearly 60 horses by Mills suggest that the regular inclusion of antacids in the diet may significantly reduce cribbing behaviour. Other strategies for reducing the risk of pH disturbance from concentrate feed include the encouragement of mastication and hence increased saliva flow through the provision of more forage, forage in nets with smaller holes to reduce the rate of intake, and compacted forage sticks which increase the time spent in the prehension phase of ingestion. Meeting the energy requirements of the diet through fat rather than carbohydrate might also be expected to produce a more stable and less acidic gastro-intestinal environment. All of these strategies appear at least anecdotally to reduce cribbing, windsucking and many cases of wood-chewing. Marsden (1999) also reports that there is an exponential inverse relationship between the proportion of time spent feeding and the proportion of time spent engaged in a range of abnormal behaviours.

The presentation of food, especially concentrate, represents a rewarding event and many horses appear to exhibit bizarre feed time rituals which have probably been conditioned through the presentation of feed. This is also a time of high arousal as food is anticipated and so frequent meal feeding or staggered feed-times across a yard may be expected to increase the stress load on a horse. The provision of an ad libitum forage based diet is therefore to be preferred when possible. Foraging devices which require the horse to work for a small amount of concentrate at any given time have been proposed for the prevention of some of these problems, but have a variable effect (Henderson & Waran 2001). They may moderate the period of high arousal associated with the delivery of food and so be efficacious for the control of pre-feeding problems, but the substantially extended time spent involved in ingesting concentrate (Winskill *et al.*, 1996) may exacerbate problems related to this component of the diet. It is popularly suggested that 'toys' help to alleviate boredom, but there is neither any evidence that horses can feel bored nor any identifiable problem which can be reliably associated with this state of mind since the evidence suggest that most stable 'vices' appear to be related to specific frustrations.

Management regimes may also be associated with an increased risk of problems such as colic, which obviously have a serious impact on the welfare of the horse. Practices which may help in the prevention of colic and which help to improve welfare in other ways include the following:

- Avoid sudden or large changes to the daily routine – including feeding and exercise.
- Feed a high quality diet comprised primarily of roughage where possible.
- Avoid feeding excessive grain and energy dense supplements. (At least half the horse's energy requirements should be supplied through hay or forage. A better guide is that twice as much energy should be supplied from roughage source than from concentrates.
- Divide daily concentrate ration into two or more smaller feedings, rather than one large one, to avoid overloading the horse's digestive tract. Hay is best fed free-choice.
- Set up a regular parasite control program with the help of your equine practitioner. Utilise fecal testing to determine its effectiveness.
- Provide exercise and/or turnout on a daily basis.
- Change the intensity/duration of an exercise regime gradually.
- Provide fresh, clean water at all times. (The only exception is when a horse is excessively hot. Then it should be given small sips of lukewarm water until recovered.)
- Avoid medications unless they are prescribed by your equine practitioner, especially pain-relief drugs (analgesics), which can cause ulcers.
- Check hay, bedding, pasture and environments for potentially toxic substances such as noxious weeds, and other indigestible foreign matter such as hay binding.
- Avoid putting feed on ground, especially in sandy soils.
- Make dietary and other management changes as gradually as possible.
- Reduce stress. Horses experiencing changes in environment or workloads are at a high risk of intestinal dysfunction.
- Pay special attention to animals when transporting them or changing their surroundings, such as at shows.
- Observe foaling mares pre- and postpartum for any signs of colic. Also watch any horses who have had a previous bout with colic.
- Maintain accurate records of management, feeding practices, and health.

5.3. OTHER REQUIREMENTS

No work has yet been published which has quantified the psychological needs of the horse. However, there have been a number of behavioural studies, which suggest the preferred requirements of horses. The interpretation of preference tests in terms of welfare is not simple since preference for one substrate over another does not imply that either is necessarily associated with suffering nor may either be good for the subject's well-being.

In addition to the bedding preferences of horse already discussed (Mills *et al.*, 2000), horses have been found to prefer a lit to a dark box stall (Houpt & Houpt 1988) and will work for access to a paddock (Houpt, personal communication). Exercise is also associated with a reduction in wood chewing (Krzak *et al.*, 1991) and possibly other behaviours of concern. Kusunose *et al.* (1985, 1987) found that yearling horses were restricted in their canter in fields of 1.5 ha but not in fields of more than 2.1 ha and that the shape of the field also affected behaviour. In the latter study (Kusunose *et al.*, 1987) suggested on the basis of the type of movement shown, that square paddocks are preferable to those proportioned 1:2 or 1:4. Whilst the limitations of these studies are accepted, it does no harm to recognise and implement their recommendations where possible, given the relative paucity of scientific knowledge available to guide decisions. At present we can only hope to adopt best practice on the basis of the areas so far recommended and careful consideration of the biological nature and expected limitations of the horse.

6. References

Alexander, F. and Hickson, J.C.D. (1970) The salivary and pancreatic secretions of the horse. In Phillipson, A.T. (ed.), *Physiology of Digestion and Metabolism in the Ruminant*. Oriel Press, Newcastle upon Tyne, UK, pp. 375–389.

Bailey, M. (1992) *Lung parasites, in Current Therapy in Equine Medicine 3*. W.B. Saunders, Philadelphia, USA, p. 332.

Bruce, J.M. (1978) Natural ventilation through openings and its application to cattle building ventilation. *J. Agricultural Engineering Research* **23**, 151–167.

Clarke, A.F. (1987) Stable environment in relation to the control of respiratory disease. In Hickman, J. (ed.), *Horse Management*, 2nd Edn. Academic Press, Orlando, USA, p. 125.

Clarke, A.F. (1993) Stable dust – threshold limiting values, exposures variables and host risk factors. *Equine Veterinary J.* **25**, 172–174.

Clarke, A.F. (1994) Stables. In Wathes, C.M. and Charles, D.R. (eds.), *Livestock Housing*. CAB International, Cambridge, UK, pp. 379–403.

Cullinane, B. (1987) *Viral respiratory disease, in Current Therapy in Equine Medicine 4*. W.B. Saunders, Philadelphia, p. 443.

Cooper, J.J. and Mills, D.S. (1997) Welfare considerations relevant to behaviour modification in domestic animals. In Mills, D.S., Heath, S.E. and Harrington, L.J. (eds.), *Proceedings of the First International Conference on Veterinary Behavioural Medicine*. Universities Federation for Animal Welfare, Potters Bar, pp. 164–173.

Cooper, J.J., MacDonald, L. and Mills, D.S. (2000) The effect of increasing visual horizons on stereotypic weaving: implications for the social housing of stabled horses. *Applied Animal Behaviour Science* **69**, 67–83.

Cuddeford, D. (1999) Why feed fibre to the performance horse today? In Harris, P.A., Gomarsall, G.M.,

Davison, H.P.B. and Green, R.E. (eds.), *Proceedings of the British Equine Veterinary Association Specialist Days on Behaviour and Nutrition.* Equine Veterinary Journal Ltd, Suffolk, UK, pp. 50–54.

Crowell-Davis, S.L., Houpt, K.A. and van Ree, J.M. (1985) Feeding and drinking behavior of mares and foals with free access to pasture and water. *J. Animal Science* **60**, 883–889.

Curtis, E.F., Raymond, S.L. and Clarke, A.F. (1996) Respirable dust and atmospheric ammonia levels in horse stalls with different ventilation rates and bedding. *Aerobiologia, International J. Aerobiology* **12**, 239–247.

Duncan, P. (1980) Time budgets of Camargue horses. 2. Adults and weaned sub-adults. *Behaviour* **72**, 27–49.

Farm Animal Welfare Council (1992) FAWC updates the five freedoms. *Veterinary Record* **131**, 357.

Harrison, R. (1964) *Animal Machines.* Vincent Stuart Ltd., London, UK.

Henderson, J.V. and Waran, N. (2001) Reducing equine stereotypies using an 'Equiball'. *Animal Welfare* **10**, 73–80.

Holcombe, S.J., Jackson, C., Gerber, V., Jefcoat, A., Berney, C., Eberhardt, S. and Robinson, N.E. (2001) Stabling is associated with airway inflammation in young Arabian horses. *Equine Veterinary J.* **33**, 244–249.

Houpt, K.A. and Houpt, T.R. (1988) Social and illumination preferences of mares. *J. Animal Science* **66**, 2159–2164.

Jackson, C.A., Berney, C., Jefcoat, A.M. and Robinson, N.E. (2000) Environment and prednisolone interactions in the treatment of recurrent airway obstruction (heaves). *Equine Veterinary J.* **32**, 432–438.

Jezierski, T. (1992) Effects of Social Isolation on heart rate in horse. In Nichelmann, M., Wierenga, H. and Braun, S. (eds.), *Proceedings of the 26th International Congress on Applied Ethology.* KTML, Berlin, pp. 387–388.

Johnson, K.G., Tyrell, J., Rowe, J.B. and Pethick, D.W. (1998) Behavioural changes in stabled horses given non-therapeutic levels of virginiamycin as Founderguard. *Equine Veterinary J.* **30**, 139–143.

Kiley-Worthington, M. (1987) *The Behaviour of Horses.* J.A. Allen, London, UK.

Krzak, W.E., Gonyou, H.W. and Lawrence, L.M. (1991) Wood chewing by stabled horses: Diurnal pattern and effects of exercise. *J. Animal Science* **69**, 1053–1058.

Kusunose, R., Hatakeyama, H., Kubo, K., Kiguchi, A., Asai, Y., Fuji, Y. and Ito, K. (1985) Behavioural studies on yearling horses in field environments 1. Effects of the field size on the behaviour of horses. *Bulletin of the Equine Research Institute* **22**, 1–7.

Kusunose, R., Hatakeyama, H., Ichikawa, F., Oki, H., Asai, Y. and Ito, K. (1987) Behavioural studies on yearling horses in field environments 3. Effects of the pasture shape on the behaviour of horses. *Bulletin of the Equine Research Institute* **23**, 1–5.

Luescher, U.A., McKeown, D.B. and Dean, H. (1998) A cross-sectional study on compulsive behaviour (stable vices) in horses. *Equine Veterinary J. Suppl.* **27**, 14–18.

Lupton, J.I. (1884) *Evils of Modern Stables.* Mayhew's Illustrated Horse Management. W.H. Allen, London, UK.

Mal, M.E., Friend, T.H., Lay, D.C., Vogelsang, S.G. and Jenkins, O.C. (1991) Behavioural responses of mares to short-term confinement and social isolation. *Applied Animal Behaviour Science* **31**, 13–24.

Marsden, M.D. (1999) Behavioural problems (stable vices). In Colahan, P.T., Merritt, A.M., Moore, J.N. and Mayhew, I.G. (eds.), *Equine Medicine and Surgery* (5th edn), Mosby, St Louis, pp. 914–931.

Mason, G.J. (1991) Stereotypies: a critical review. *Animal Behaviour* **41**, 1015–1037.

McAfee, L.M., Mills, D.S. and Cooper, J.J. (in press) The use of mirrors for the control of stereotypic weaving behaviour in the stabled horse. *Applied Animal Behaviour Science.*

McBride, G.E., Christopherson, R.J. and Sauer, W.C. (1983) Metabolic responses of horses to temperature stress. *Journal of Animal Science* **57**, 175.

McBride, S.D. and Cuddeford, D. (2001) The putative welfare-reducing effects of preventing equine stereotypic behaviour. *Animal Welfare* **10**, 173–189.

McBride, S.D. and Long, L. (2001) Management of horses showing stereotypic behaviour, owner perception and the implications for welfare. *Veterinary Record* **148**, 799–802.

McGreevy, P.D., Cripps, P.J., French, N.P., Green, L.E. and Nicol, C.J. (1995) Management factors associated with stereotypic and redirected behaviour in the Thoroughbred horse. *Equine Veterinary J.* **27**, 86–91.

McGreevy, P.D. and Nicol C.J. (1998) The effect of short term prevention on the subsequent rate of crib-biting in Thoroughbred horses. *Equine Veterinary J. Suppl.* **27**, 30–34.

Mills, D.S. (1999) The origin and development of behavioural problems in the horse. In Harris, P.A., Gomarsall, G.M., Davison, H.P.B. and Green, R.E. (eds.), *Proceedings of the British Equine Veterinary Association Specialist Days on Behaviour and Nutrition.* Equine Veterinary Journal Ltd., Suffolk, UK, pp. 17–21.

Mills, D.S. and Davenport, K. (in press) The effect of a neighbouring conspecific versus the use of a mirror for the control of stereotypic weaving behaviour in the stabled horse. *Animal Science.*

Mills, D.S., Eckley, S. and Cooper, J.J. (2000) Thoroughbred bedding preferences, associated behaviour differences and their implications for equine welfare. *Animal Science* **70**, 95–106.

Mills, D.S. and Nankervis, K.J. (1999) *Equine Behaviour: Principles and Practice.* Blackwell Science, Oxford.

Murray, M.J., Schusser, G.F., Pipers, F.S. and Gross, S.J. (1996) Factors associated with gastric lesions in Thoroughbred racehorses. *Equine Veterinary J.* **28**, 368–374.

Nicol, C.J. (1999) Understanding equine stereotypies. *Equine Veterinary J. Suppl.* **28**, 20–25.

Ogilvie-Graham, T.S. (1994) *Time Budget Studies in Stalled Horses.* DVM&S Thesis, University of Edinburgh, Edinburgh, UK.

Pagan, J.D. (1997) Gastric ulcers in horses: A widespread but manageable disease. *World Equine Veterinary Review* **2**, 28–30.

Racklyeft, D.J. and Love, D.N. (1990) Influence of Head posture on the respiratory tract of healthy horses. *Australian Veterinary J.* **67**, 402–405.

Ralston, S.L., van den Broek, G. and Baile, C.A. (1979) Feed intake patterns and associated blood glucose, free fatty acid and insulin changes in ponies. *J. Animal Science* **49**, 838–845.

Raymond, S.L. and Clarke, A.F. (1997) Equine Respiratory Health, World Equine Veterinary Review. *The J. the World Equine Health Network* **2**, 41–44.

Raymond, S.L., Curtis, E.F. and Clarke, A.F. (1994) Comparative dust challenges faced by horses when fed alfalfa cubes or hay. *Equine Practice* **16**, 42–47.

Raymond, S.L., Curtis, E.F., Winfield, L.M. and Clarke, A.F. (1997) A comparison of respirable particles associated with various forage products for horses. *Equine Practice* **19**, 23–26.

Raymond, S.L., Heiskanan, M.L., Smith, T.K., Reiman, M., Laitinen, S. and Clarke, A.F. (2000) An investigation of the Concentrations of Fusarium mycotoxins and the degree of mould contamination of Ontario field-dried hay. *J. Equine Veterinary Science* **20**, 616–621.

Rees, L. (1984) *The Horse's Mind.* Stanley Paul, London, UK.

Rolls, E.T. (1990) A theory of emotion and its application to understanding the neural basis of emotion. *Cognition and Emotion* **4**, 161–190.

Rowe, J.B., Pethick, D.W. and Lees, M.J. (1994) Prevention of acidosis and laminitis associated with grain feeding in horses. *J. Nutrition* **124**, 2742–2744.

Sainsbury, D.W.B. (1987) Housing the Horse. In Hickman J. (ed.), *Horse Management.* Academic Press, London.

Schatzmann, U. (1998) Winter pasturing of sport horses in Switzerland – an experimental study. *Equine Veterinary J. Suppl.* **27**, 53–54.

Tyler, S.J. (1972) The behaviour and social organisation of the New Forest ponies. *Animal Behaviour Monographs* **5**, 2.

Welford, D., Mills, D., Murphy, K. and Marlin, D. (1999) The effect of changes in management scheduling on water intake by the Thoroughbred horse. *Equine Veterinary J. Suppl.* **28**, 71–72.

Wilson, W.D. (1997) *Foal pneumonia in Current Therapy in Equine Medicine 4.* W.B. Saunders, Philadelphia, USA, p. 612.

Winskill, L.C., Young, R.J., Channing, C.E., Hurley, J. and Waran, N.K. (1996) The effect of a foraging device (the modified 'Edinburgh Foodball') on the behaviour of the stabled horse. *Applied Animal Behaviour Science* **48**, 25–35.

Chapter 5

STEREOTYPIC BEHAVIOUR IN THE STABLED HORSE: CAUSES, EFFECTS AND PREVENTION WITHOUT COMPROMISING HORSE WELFARE

J. COOPER

Animal Behaviour, Cognition and Welfare Research Group, University of Lincoln, Lincolnshire School of Agriculture, Caythorpe Campus, Lincolnshire NG32 3EP, UK

P. McGREEVY

Faculty of Veterinary Science, Gunn Building (B19), Regimental Crescent, University of Sydney, NSW 2006, Australia

Abstract. Apparently functionless, repetitive behaviour in horses, such as weaving or crib-biting has been difficult to explain for behavioural scientists, horse owners and veterinarians alike. Traditionally activities such as these have been classed amongst the broad descriptor of undesirable stable vices and treatment has centred on prevention of the behaviours per se rather than addressing their underlying causes. In contrast, welfare scientists have described such activities as apparently abnormal stereotypies, claiming they are indicative of poor welfare, citing negative emotions such as boredom, frustration or aversion in the stable environment and even suggesting prevention of the activities alone can lead to increased distress. Our understanding of equine stereotypies has advanced significantly in recent years with epidemiological, developmental and experimental studies identifying those factors closely associated with the performance of stereotypies in stabled horses. These have allowed the development of new treatments based on removing the causal factors, improving the horses' social and nutritional environment, re-training of horses and their owners and redirection of the activities to less harmful forms. Repetitive activities conventionally seen as undesirable responses to the stable environment, their causal basis and the effectiveness of different approaches to treatment are discussed, both in terms of reducing the behaviour and improving the horse's quality of life.

1. Introduction: Investigating Stereotypy in the Horse

In the stable, there is considerable deviation from the behavioural patterns of the wild or free-ranging horse. Stereotypic patterns of behaviour, such as weaving, crib-biting and box-walking, are particularly associated with stabling, affecting between 10 and 40% of stabled horses (Nicol 1999a). These are often described as 'abnormal' as they are; rarely observed in free-ranging horses, difficult to explain in functional terms, undesirable to horses owners, and because they can lead to or are caused by, welfare problems for the stabled horse (Cooper & Mason 1998). In stable management, stereotypies are traditionally classed with a wider category of 'vices' that are considered undesirable to people (*e.g.* Houpt 1982; Luescher *et al.*, 1998). These terms are, however, unhelpful when considering the welfare of the horse, as they emphasise the inconvenience of the behaviours to humans, rather

N. Waran (ed.), The Welfare of Horses, 99–124.
© 2002 *Kluwer Academic Publishers. Printed in the Netherlands.*

than simply describing the observable events or seeking to understand their root causes and functional significance through objective investigation.

The understanding of causes and effects of equine stereotypies are investigated in epidemiological studies by McGreevy et al. (1995a, b), Luescher et al. (1998) and Redbo et al. (1998), which have been reviewed by Nicol (1999a). These consistently relate the incidence of stereotypies to a number of management factors including the feeding of concentrates with little access to fibre and social isolation. On their own, surveys can however only show a correlation between behaviour and management practises, so empirical studies are required to investigate the causal relationship between these environmental factors and the development and perseverance of stereotypy.

Empirical studies aim to investigate the effects of varying specific factors on the incidence of stereotypy under controlled conditions. Equine ethologists have investigated the effectiveness of traditional preventative measures such as weaving grills (McBride & Cuddeford 2001), anti-cribbing devices (McGreevy & Nicol 1998a) and the effects of pharmacological intervention (Dodman et al., 1987). Recent studies have also measured the horses' physiological distress responses such as heart rate or adreno-cortical activity (Broom & Johnson 1993) to test if there are any underlying effects of treatment on horse welfare (Lebelt et al., 1998; McGreevy & Nicol 1998b, c; McBride & Cuddeford 2001). Generally these studies have found that preventative measures alone increase distress, suggesting a compromise of horse welfare. A number of alternative approaches have also been investigated, including foraging devices (Winskill et al., 1996), feed additives such as fibre and anti-acids (Johnson et al., 1998; Nicol et al., 2001), increasing social contact (Cooper et al., 2000) and even mirrors (Mills & Davenport, in press; McAfee et al., in press). Initial results from such studies are encouraging with significant reductions in stereotypic behaviour with no apparent compromise of horse welfare.

The final pieces in the jigsaw are developmental studies of stereotypy. It is well known that many stereotypic activities change as they develop, becoming more repetitive, more divergent from their original root behavioural patterns and often harder to disrupt both behaviourally (Cooper et al., 1996) and pharmacologically (see Mason 1991, for review). Empirical studies alone, as they are only a snapshot of the occurrence of the behaviour, may lead to mis-interpretation if the horse has been performing the activity for long periods of time. Developmental studies, in contrast, track changes in behaviour through time, and are particularly useful in investigating the relationship between stereotypy and their effects. Horses are, however, expensive to maintain and equine stereotypies may arise gradually so developmental studies are practically and financially difficult to perform. Nevertheless, the results of such long-term studies are now becoming available (Nicol 1999a; Waters et al., in press) and show, for example relationships between weaning practice and both incidence and type of stereotypy.

2. Activities Described as Equine Stereotypies

Two locomotor activities are classically described as stereotypic; weaving and box-walking. Weaving has been defined as the lateral swaying of the head over the stable door or some other barrier (Mills & Nankervis 1999) (see Figure 1). The activity may also involve the swaying of the rest of the body, including the shoulders, and picking up the front legs. In stables with weaving grills a similar activity called treading (Kiley-Worthington 1987) may be observed, involving swaying of the body or alternative lifting of the forelegs, but without the swaying of the head and neck. Box-walking has been defined as the pacing of a fixed route around the stable (Kiley-Worthington 1983). Typically, a circular route is traced but in larger stables or in the field, horses may trace a 'figure of eight' shaped route.

In addition, several other repetitive body or limb movements such as nodding, door-kicking and pawing can also be described as stereotypic. Nodding involves the vertical movement of the head and neck typically whilst held above the stable door or other barriers (Cooper *et al.*, 2000) which is morphologically distinct from weaving. Nodding can also be readily distinguished from head-shaking and aggressive head movements as it involves the repetition of the head movement, though

Figure 1. A horse weaving over a stable door [Courtesy of Natalie Waran].

it may have developed from such activities as a conditioned response to stable management practices. Door kicking involves kicking the stable door, walls or other stable furnishings with the fore-legs. These and similar activities such as pawing the ground are commonly seen in stabled horses prior to feeding or other potentially stimulating periods of the day (Kiley-Worthington 1983; Mills & Nankervis 1999).

Several oral activities have classically been described as stereotypic, including crib-biting, wind-sucking, grasping and wood-chewing (Owen 1982). Crib-biting involves the grasping of a surface (usually horizontal) in the teeth (McGreevy & Nicol 1998a) and the apparent engulfing of air (see Figure 2). Typical cribbing surfaces include the top of the stable door, shelves or horizontal lips on the stable wall or the edges of feed or water buckets. Wind-sucking involves the same contraction of neck muscles and apparent engulfing of air, but without grasping. Both cribbing and wind-sucking can be characterised in behavioural observations as 'an apparent sharp intake of breath' and are often accompanied by an audible 'grunt'. Wind-sucking is often also called aerophagia (air swallowing) in scientific literature (*e.g.* Karlander *et al.*, 1965; Baker & Kear-Colwell 1974; Kuusaari 1983). This is, however, a misnomer as tracing of the air movements in the respiratory tracts of wind sucking horses reveal that little or no air is swallowed (McGreevy *et al.*, 1995c).

Grasping where the horse holds or bites stable fittings without the apparent air-engulfing can also be described as stereotypic (Owen 1982) when the horse

Figure 2. A horse performing crib-biting in a racing stables in India (Courtesy of Natalie Waran).

repeatedly grasps the same location. Finally, wood-chewing involves the grasping, stripping and apparent ingestion of wooden surfaces in the stable, such as the top of the door or edges to stable walls (Krzak *et al.*, 1991), though it can also be seen in free-ranging horses on fence posts and trees. Wood-chewing and grasping can be confused if the cribbed surface is wooden, and careful attention needs to be made to establish if the horse is ingesting wood (lignophagia) when discriminating between the two. Wood-chewing is considered a behavioural problem as it leads to damage to stable fittings and there is concern that ingestion of wooden splinters may harm the horse (Green & Tong 1988).

Several other activities that are considered undesirable by stable managers may be better described as redirected behaviour or even considered the normal learnt response to stable management. An example is bed-eating, which involves the chewing and ingestion of bedding substrates such as straw, paper or shavings (Mills *et al.*, 2000). It is often accompanied by nosing and sifting for hard feed or forage that has fallen to the floor and closely resembles the natural foraging patterns of feeding horses. It commonly follows the feeding of concentrates and may represent attempts to find spilt food or perseverance of feeding behaviour following a small meal or even attempts to balance the low fibre concentrated meal with more fibrous material. Bed-eating is more common in horses that do not have access to high fibre forages such as hay and it is more common in horses on straw bedding than in horses housed on rubber, paper or shavings (Mills *et al.*, 2000). Horses on straw also show few oral stereotypies such as crib-biting (McGreevy *et al.*, 1995a; Mills *et al.*, 2000). Bed-eating is not considered a stereotypy in this chapter as it appears to be part of the horses' repertoire of feeding behaviour and diet selection in the stable.

A number of other mouth movements that are not conventionally described as 'vices' may, however, be described as stereotypic. These include repetitive licking of non-food items such as the stable wall, floor, the sides of food buckets or other stable furniture. Normally licking would have a function such as picking up spilt feed, water or attempts to intake trace nutrients, but where the behaviour is repetitive and focussed on a single location, then it may conform to the definition of stereotypy. Other repetitive activities seen in horses include sham-chewing or tooth-grinding where the horse performs repetitive tongue, mouth or jaw movements without any obvious food substrate in the mouth. These activities are frequently observed around the time of feeding concentrates (Willard *et al.*, 1977) and may represent the expression of feeding motivation, either in anticipation of feed delivery or following the consumption of a concentrate meal. These mouth movements resemble activities that are conventionally described as stereotypic in other species, such as vacuum chewing in pigs or tongue-rolling in calves. Such activities in horses have rarely been considered a problem either by horse owners or horse observers so few studies have investigated their incidence, causation or treatment, though it is probable that they ontogeny is related to that of other oral stereotypies.

Another activity commonly seen in stabled horses that can be described as stereotypic is the repetitive adoption of facial expressions ('face pulling') in response to specific cues. These expressions can involve lip-curling, teeth baring or apparently agonistic responses such as attempts to bite and actual bites. They often arise in stabled horses at feeding time or accompany interactions with humans or other horses. The handler's interpretation of these social signals (sometimes labelled as 'nippy', 'stroppy' or 'pulling faces') may help to identify individuals and their characteristics on the stable yard, but also represent misleading anthropomorphism of horses' behaviour (Mills 1998). Whilst they share a degree of morphological similarity as they involve changes in facial characteristics in response to environmental challenges, it is difficult to classify these activities as a cohesive category of behaviour, as they are can be unique to individuals and probably develop from a variety of root behaviours. Again little research has been carried out into this diverse group of responses, though it is likely that they originated as part of the horse's complex communication and social signalling behaviour and may have been reinforced by feeding or the social responses of people or other horses.

2.1. THE CAUSES OF EQUINE STEREOTYPY

Stables differ from the free-ranging environment in a number of factors including space, nutritional environment, social environment, types of environmental substrates and the ability to make controlled environmental choices. These factors may individually or in combination contribute to the development of stereotypies. There may, for example, be an overall deficit in environmental stimulation, which might lead to emotional states such as boredom, deprivation or frustration. Specific factors that differ include the changes in the nutritional environment and choice of feeds. Horses are naturally foragers who consume large quantities of fibrous feed (Waring 1983; Harris 1999), whereas many stabled horses, particularly performance horses have carefully controlled diets, with a high reliance on concentrated feeds. Furthermore, the stabled horse has little opportunity to control the timing of feeding and other procedures such as turning out and exercise. This routine may not only contribute to the amount of stereotypy but also to their timing. The social environment also conventionally differs between the stable and the field, where horses may have more choice over social spacing and companions. Finally, the rearing environment of the animals is also likely to influence the development of stereotypy, either by directly preventing specific activities that are re-directed into stereotypy or more generally affecting the development of the horse's behaviour and its ability to learn new adoptive strategies to challenges faced in later life.

3. General Factors Associated with Housing: Boredom and Frustration

It is often claimed that stereotypic behaviour is a response to boredom in captive animals such as stabled horses (Kiley-Worthington 1987), and it occupies the time that would otherwise be engaged in other functionally significant activities such as grazing, social interactions, or predator avoidance. This argument is tempting, as many of the environmental challenges faced by free-ranging horses are no longer an issue in the stable as they are provided by the management regime. In the absence of other environmental challenges stereotypy may either have a time filling function (McFarland 1989) or may compensate for low environmental stimulation by providing sensory substitutes (Wiepkema 1985).

However, in the majority of stereotypic activities observed in captive animals, including stabled horses, there is little evidence to support this hypothesis. During quiet times of the day, the majority of horses spend their time either dozing or foraging and stereotypy is rare. Bouts of stereotypy tend to be focussed on periods of high environmental stimulation such as feeding time and prior to turning out (Johnson *et al.*, 1998; Cooper *et al.*, 2000; Henderson & Waran 2001) rather than periods of low environmental activity. This occurs on quiet yards where episodes of environmental stimulation are confined to feeding times and turning out and on busy yards where there is a high level of human disturbance as well as the stimulation of exercise and feeding times (McAfee *et al.*, 2002).

Low environmental stimulation may be indirectly related to stereotypy, not as a means of increasing environmental stimulation, but as a consequence of low behavioural competition. Contemporary models of animal motivation (*e.g.* Toates 1981; McFarland 1989) state that an animal's decision making process is based on a number of competing behavioural systems, such as feeding or predator avoidance. The importance of the activities associated with these systems (and consequently the likelihood of performance) will depend on the animal's internal state (*e.g.* hunger) and external cues (*e.g.* presence of food, absence of predators). As these behavioural systems compete for expression, high priority activities will exclude or even suppress the expression of low priority activities. In this control system, the removal of competing cues, for example predatory stimuli, not only reduces the motivation to perform related anti-predatory activities, but also removes the inhibition of other responses that would occur with performance of these anti-predatory responses. Consequently stereotypies are more common in environments with low environmental complexity; firstly because there is the absence of appropriate substrates for expression, secondly because the range of alternative substrates is limited and thirdly because there is little behavioural competition to inhibit their performance (Cooper *et al.*, 1996).

Another common general explanation of stereotypy is 'behavioural frustration' where repeated activities are derived from the perseverance of highly motivated activities that can not be adequately expressed in the captive environment (Hughes & Duncan 1988). This explanation is supported by the form and timing of the majority of stereotypies in domestic animals. For example, pre- and post-feeding stereotypies are common in a number of animals (Mason & Mendl 1997), including dry sows who perform sham rooting, bar-biting and sham chewing at feeding time and mink which somersault and pace prior to feeding. In these cases the stereotypy may either represent the perseverance of the activity in the absence of a satisfactory end point or the redirection of the activity to an apparently functionally irrelevant alternative. In horses a number of highly motivated activities that cannot be adequately expressed in conventional stables may lead to stereotypy. These include the prevention of locomotor activity and exercise, restrictions on diet selection and foraging, and imitations on both the social environment and predator avoidance activities (Houpt & McDonnell 1993; Cooper & Mason 1998; Nicol 1999b).

3.1. CONCENTRATED FEEDING

A number of authors have implicated the feeding of concentrates as a significant factor in the development of stereotypies (Kiley-Worthington 1987; Houpt & McDonnell 1993; Marsden 1993) and these suggestions are supported by empirical studies (Willard *et al.*, 1977; Johnson *et al.*, 1998). The underlying rationale is that horses are naturally free ranging grazing herbivores that spend much of their time feeding to maintain balanced energy and nutrient intake. In the stable, the requirement to forage has been removed as horses are provided with a balanced diet. Although such diets are theoretically formulated to meet all the horse's dietary needs, they may not meet the horse's behavioural needs. Conventional stable rations come in a concentrated form which take little time to eat and contain considerably less bulk or dietary fibre than the diet the horse has evolved to consume. Under these circumstances, horses may still be motivated to forage after the concentrate has been eaten. If, for example feeding time is controlled by gut fill or chewing time (Cooper & Mason 1998), then the horse may only be able to express the motivation to feed as an analogue such as crib-biting, sham-chewing or wood-chewing. In addition concentrated feeding may lead to digestive problems as dietary fibre is involved not only in gut-fill but also in gut transport and buffering of digestive systems (Harris 1999; Nicol 1999b). In particular, grain-based concentrate feeds may lead to harmful levels of gut acidity (Rowe *et al.*, 1994; Murray and Eichorn 1996) and the high levels of repetitive oral activity (included grasping, crib-biting, licking and sham-chewing) associated with concentrated feeding may be a response to gut conditions (Nicol *et al.*, 2001).

Providing less concentrated feed or more fibrous forages (*e.g.* hay or haylage) may reduce the intensity or frequency of oral stereotypy by allowing foraging of more appropriate substrates. This prediction is supported by both epidemiological and empirical evidence. McGreevy *et al.* (1995a) found that low forage was a major risk factor in the incidence of stereotypies, whilst Marsden (1993) has reported that providing forage (soaked hay) significantly reduces oral stereotypies post feeding. Furthermore, the incidence of post feeding oral stereotypies is significantly higher in horses that do not have access to hay at feeding time, compared with horses that receive hay at the same time as their concentrate feed (unpublished, Cooper *et al.*). The relationship between oral stereotypies and forage is probably best illustrated by the studies of McGreevy and Nicol (1998c) on the prevention of cribbing and grasping. They found that removing a favoured cribbing surface from a horse's stable significantly increased plasma cortisol (suggesting a physiological response to frustration) unless the horses had access to fresh hay. This not only illustrates the dangers of preventing stereotypy, but also that oral stereotypies and foraging may act as partial substitutes for each other if they share the same motivational root.

Another means of addressing the horse's motivation to forage without providing

Figure 3. A horse feeding using an EquiballTM (Courtesy of Natalie Waran).

high fibre feeds, is to use foraging devices such as the EquiballTM (see Figure 3). The rationale behind such devices is that not only does the horse take longer to eat its ration as the feed trickles from the device, but also it allows a mode of expression for the horse's foraging behaviour. Empirical studies indicate that such devices increase time spent 'foraging' in stabled horses (Winskill *et al.*, 1996), but they may not reduce oral stereotypies, as these activities are spread over a longer period of time. Foraging devices appear to be more effective at reducing loco-motor stereotypies such as box-walking or pre-feeding stereotypies such as weaving (Henderson & Waran 2001). Whether such devices are effective by removing the underlying causes of the stereotypy, providing a more acceptable substitute or whether they simply represent the extinction of a learnt response will be explored later in this chapter.

3.2. TURNING OUT AND EXERCISE

Another factor that may be associated with the performance of stereotypic behav-iour, is the amount of exercise or 'turn-out time' that horses receive. It has been suggested that unless stabled horses have the opportunity to perform kinetic activity outside the stable, they may respond by performing undesirable locomotor activi-ties in the stable such as weaving or box-walking (McGreevy 1996). There is however little evidence that exercise routine has a consistent effect on incidence of stereotypy in the stable (Marsden 1993) and exercise routine was not identified as a risk factor in an epidemiological study (McGreevy *et al.*, 1995a). The evidence for a relationship between exercise routine and stereotypy is therefore equivocal. Evidence does however exist to suggest that turning out or exercise acts as a focus for the expression of stereotypic behaviour, with Cooper *et al.* (2001) finding a high incidence of weaving immediately prior to turning out and Waran and Campbell (unpublished data) finding that the period post-exercise is associated with the performance of cribbing in thoroughbred horses undergoing training.

Stereotypic behaviour itself is rarely observed in the field either because the horses are not exposed to the cues that stimulate stereotypies or if they are, the underlying motivation can be more appropriately satisfied. Some horses will nevertheless perform stereotypic behaviour in a relatively 'free-living' environment (for example weaving over a field gate or cribbing on fence posts). This is seen both in older horses with prior experience of stables who may persevere with the behaviour in other environments and in young foals. McGreevy (1996) identified a small number of foals that performed crib-biting on fence posts and rails prior to weaning. These activities did not appear to be related to any deficiencies in maternal care as the dams were described as good mothers with good milk supply. In a more detailed study by Waters (2002) concentrating on the development of stereotypic activities, these pre-weaning stereotypies were only found in foals that received grain-based dietary supplements prior to weaning, which supports the relationship between diet and oral stereotypy found in older stabled horses (Johnstone *et al.*, 1998; Nicol *et al.*, 2001).

In the cases of older horses with experience of stabling, the stereotypy may become emancipated from its original causes, either because of the endogenous reinforcement of the activity or failure to extinguish the response in the field. Emancipation is the process by which behaviours become independent of their original causal factors (Cooper *et al.*, 1996) and has been widely accepted to occur with stereotypic patterns of behaviour. Instead of reflecting current problems, an established stereotypy may be likened to a neurological 'scar' that marks a legacy from past conditions (Mason 1991). Empirical evidence for emancipation is, however, rare and confined to studies of laboratory animals (*e.g.* Cooper *et al.*, 1996; Powell *et al.*, 2000). In horses the emergence of stereotypic behaviour seems to be reduced in foals that are given an enriched environment, but the same effect has not been observed in older animals (Nicol 1999a). This suggests that foals are particularly reactive to stimuli that elicit stereotypies. It is generally thought that, with age, equine stereotypies are elicited by a wider set of stimuli than in early development and, further, that they become relatively resistant to normal control mechanisms. This is why older horses are commonly seen stereotyping in environments that appear to support high welfare and may explain some of the variation in effectiveness of treatments or preventative measures. In these animals, the stereotypic activities may also have become part of their repertoire of learned behaviour, activated as a conditioned response in anticipation of a particular eliciting event (*e.g.* a concentrate meal, removal of a companion).

3.3. ROUTINE AND CONDITIONING

Many stable yards follow relatively set routines. Horses may, for example be fed, exercised and returned to the stable at roughly the same time each day. Within these routines, there is potential for horses not only to learn when key events occur during the day (*e.g.* feeding time), but also what events precede and reliably predict these events (*e.g.* certain people on the yard, activity in the feed room). Under these conditions there is considerable potential for horses to learn to associate their actions, for example, kicking the door, with particular rewarding events that follow them. For example, a horse that kicks the door may be rewarded by receiving additional attention or by being fed earlier within a yard, or it may associate its door-kicking actions with feeding time. Horses may naturally perform a number of appetitive or anticipatory activities prior to feeding, such as approaching the feed or pawing the ground. In free-ranging conditions these may lead to acquisition of the feed, whereas in the stable, the actual feeding is dependent on the carer and consequently beyond the horses' behavioural control. If, however, these incomplete anticipatory actions are reliably followed by rewarding events such as feeding or turning out, then the horses will nevertheless learn to associate performance of the anticipatory action with its apparent rewarding consequences.

There is good evidence that stereotypy has become a conditioned response to events on the equine yard. Activities such as weaving, nodding and box-walking are most prevalent before feeding time (Winskill *et al.*, 1996; Cooper *et al.*, 2000). The event of turning out can also be a cue that initiates the performance of nodding and weaving (Cooper *et al.*, 2000). This may either be a conditioned response to a desirable outcome such as leaving the stable, or the expression of species typical anticipatory activities in an unusual form such as attempts to socially interact with other horses being led to the field (Nicol 1999b; Cooper *et al.*, 2000).

Changing the horses' routine also suggests that certain horses have learnt to stereotype prior to feeding. For example, changing the feeding time or changing the cues that pre-cede feeding dramatically reduces pre-feeding stereotypy (unpublished, Cooper *et al.*), but not post-feeding stereotypy. This may explain the effect of trickle feeding devices such as the Equiball™ (Winskill *et al.*, 1996; Henderson & Waran 2001) *i.e.* they reduce pre-feeding stereotypy by removing pre-feeding cues rather than acting as an alternative means of expression for pre-feeding motivation. If this is true, then changing husbandry routine may be an effective treatment of stereotypy. However, there are difficulties with these approaches. Firstly, there is the danger of an initial increase in stereotypy following the dis-association of environmental cues and reward. This is typical of extinction of conditioned response, prior to the actual loss of association and presents a problem in convincing horse managers about the effectiveness of a change in husbandry. Secondly, any decrease in stereotypy following change in routine may be short lived as the horses learn new associations that predict feeding in the new routine. Consequently in the longer term stereotypy may return to its former levels and the husbandry routine may need to be continually changed to maintain the effect.

3.4. SOCIAL ENVIRONMENT AND OBSERVATIONAL LEARNING

Horses are social animals and the horses' social environment is often overlooked in stable management. Conventionally horses are singly housed in stables for ease of management. Whilst this may have benefits in limiting opportunities for negative social interactions such as biting or kicking, these may well be outweighed by prevention of positive social interactions. It is also believed that social housing can lead to the social facilitation, copying or mimicry of stereotypy. This could occur where introducing a horse that reliably performed a stereotypy, such as weaving or wind-sucking, caused other previously non-stereotypic horses in the yard to take up the activity, rather like an infectious disease. For this reason many horse owners avoid mixing stereotypic with non-stereotypic horses (McBride & Long 2001) and even visually isolate stereotypic horses. Whilst there is some empirical evidence of social facilitation of stereotypy in other species such as voles (Cooper & Nicol 1994), there is no such evidence from horses where accounts of copying are largely anecdotal and subject to scepticism (Cooper & Mason 1998; Nicol 1999b).

In contrast both epidemiological and empirical studies of stereotypy have suggested that enhancing the horses' social environment actually reduced the incidence of stereotypy. For example, McGreevy et al. (1995a) found that stereotypy was less common on large yards where horses had visual contact with a large number of other horses. The low incidence of stereotypy in these yards may be related to a number of factors (e.g. increased yard activity compared with smaller yards) but within these yards increased visual contact with other horses was also a factor in lowering risk of stereotypy. In empirical studies, allowing close visual and tactile contact with the neighbouring horse (directly through a grill between stables, as opposed to when the horses happened to have their heads out of the stable door) significantly reduced weaving and nodding relative to the conventional stable (Cooper et al., 2000).

This increased close social contact may help explain the low incidence of stereotypy in stall-tied horses such as those of the household cavalry who have restricted locomotion but greater opportunity for social interaction than conventionally stabled horses (Marsden 1993). It may, however, be inconvenient or impractical to socially house all horses, due to risk of infection, undesirable social interactions, or just the cost of maintaining an additional horse. A simple alternative may be the use of stable mirrors, which appear to have a similar effect to social contact in both short (Davenport et al., 2001) and long term (McAfee et al., 2002) studies. Whether horses with mirrors 'see' another horse or are merely distracted by the movement is not clear, but whatever the horses see, appears to reduce weaving.

3.5. WEANING AND REARING ENVIRONMENT

Researchers at the University of Bristol recently completed a four-year study of 225 thoroughbred and part-thoroughbred horses weaned between 3 and 9 months of age (Nicol 1999a; Waters et al., in press). The study focussed on weaning practice of warm-blooded horses, selecting foals that were exposed to one of three common weaning strategies, stable weaning, barn weaning and paddock weaning and following the development of stereotypies in later life. Stable weaning involved removal of the foal from the mother and housing it in isolation in a stable or box. Barn weaning involved a number of foals being weaned and group housed without the mothers in a large arena. Paddock weaning involved housing mothers and foals in a large enclosure (e.g. a field) then removal of one mare at a time until only foals remained in the field. The study consisted of direct observations of foals in the months around weaning, and tracking the incidence of stereotypy in subsequent years.

The study found the incidence of oral stereotypies to be much higher than expected from other epidemiological studies (Nicol 1999a), with 10.5% of foals developing crib-biting and 30.3% of foals developing wood-chewing. This could be because direct observation was used to record stereotypy, whilst most epidemiological studies rely on owner based surveys alone. Alternatively, horses with high incidence of oral stereotypies may be culled from the adult population or oral stereotypies may simply be less common in older horses. The incidence of weaving (4.6% of horses) and box-walking (2.3% of horses) were consistent with other epidemiological studies of adult horses. These locomotor activities tended to arise later than oral stereotypies in the study population at a median age of 60 weeks for weaving and 64 weeks for box-walking compared with 30 weeks for wood-chewing and 20 weeks for crib-biting.

Weaning practice was a major factor in both the incidence and the type of stereotypy. Stereotypic behaviour was least common in paddock weaned foals, with little overt disturbance in behaviour beyond the initial response to removal of the mare. This supports the studies of Houpt et al. (1984) and McCall et al. (1985), both of which found group weaning to cause less stereotypy in horses than weaning in isolation. Box-walking was most commonly observed in stable-weaned foals following separation from the mare, whereas barn-weaned foals commonly nosed and mouthed con-specifics, an activity that may represent redirection of suckling (Nicol 1999a). Both stable- and barn-weaned horses performed more wood-chewing in later years than paddock weaned horses and the development of crib biting appeared to be associated with the feeding of concentrates following weaning (Waters et al., in press). In contrast, weaving did not appear to be related to weaning practice and many horses start to weave when they are about one year old, after they have been sold from the stud to new homes.

These epidemiological, experimental and developmental studies, therefore, give an insight into the causal factors underlying the development of stereotypy and their continued performance in later life. Having described the activities commonly labelled stereotypic in horses and their causes, the remainder of the chapter will deal with the consequences of different approaches to preventing stereotypy and good preventative management practises without compromising the horse's quality of life.

4. Consequences of stereotypies and problems with traditional and current remedies

4.1. WHY TREAT EQUINE STEREOTYPIES?

The reasons for seeking to reduce the incidence of stereotypic behaviour in horses can broadly be divided into two areas. These are, firstly the undesirable physical or behavioural consequences of the activities for the horse and secondly the attitudes of horse owners. Reasons that fall within the former area offer good grounds for seeking to reduce stereotypy so long as the preventative measures are effective and so long as the costs of the treatment to the horse do not outweigh the benefits of prevention. Owner attitudes may seem to be not a good reason for treatment, however, they are a major factor in horse husbandry and ethologists and veterinarians need to be sympathetic to the culture of horse owners when delivering practical advice on managing stereotypy (McBride & Long 2001).

Performing stereotypies incurs costs on the horse in terms of time and energy. If stereotypies are merely a time-filling activity, then there should be little concern about an increase in time spent on their performance. If, however, performing stereotypy impinges on other activities (such as eating or resting) then it may interfere with the horse's ability to respond to its environment. All non-feeding stereotypic behaviours waste energy, so there is usually a reduced ability to sustain bodyweight in any horse that spends a significant amount of time in such activities rather than feeding and resting (McGreevy et al., 2001). Weaving is also thought to cause excessive wear and tear on the hooves and the musculo-skeletal system. Box-walking in a single direction can cause lateralised atrophy and hypertrophy of the lumbar musculature. Self-mutilation can lead to bite wounds and secondary infections.

Crib-biting has been associated with digestive disorders, tooth erosion and failure to thrive. Owen (1982) notes that crib-biting leads to flatulence and chronic colic. Unless specially modified sites are provided, tooth wear is an inevitable consequence of crib-biting. While such tooth wear is unlikely to be associated with pain since nerves are not exposed, it can result in difficulty in grasping and chewing forage, and this may ultimately result in loss of body condition. Weight loss may also occur in crib-biters if planes of nutrition are critical, because these horses spend more time and energy performing the behaviour and less time resting and nourishing themselves (McGreevy 1995).

Although the link between wind-sucking and/or crib-biting and digestive disorders such as colic has yet to be proved, it is commonly assumed by both horse owners and veterinarians (Ritzberger-Matter & Kaegi 1998). This may be because the incidence of colic is low in both stereotypic and non-stereotypic populations of horses, but that the risk of colic is cited when justifying surgical or physical prevention of crib-biting and 'wind-sucking'. One potential explanation linking colic to the ingestion of air has been challenged by radiographic evidence that indicates the ingestion of minimal volumes of gas during crib-biting (McGreevy et al., 1995c). Even if a relationship between digestive disorders and stereotypy were to be found, this would not mean stereotypies cause these problems as they may be correlational, with both symptoms sharing another root cause or it may even be that the gut disorder leads to stereotypy. For example, if low fibre diets cause high gut acidity, then increased oral activity may be a means of lowering acidity by saliva production (Nicol 1999b). This is supported by the finding that crib-biting horses experience more gut erosion and mucosal erosion in comparison to non-cribbers (Nicol et al., 2001).

If the benefits to the horse of allowing the crib-biting outweigh the concerns felt by owners then it could be argued that allowing continued performance is desirable to outright prevention (McGreevy & Nicol 1998a). This could be achieved by allowing affected horses to crib-bite, but to provide with adequate foraging opportunities. The damage done to incisors during a lifetime performing this behaviour could also be minimised by the provision of cushioned cribbing-bars throughout the horse's environment. However, this would meet with opposition from those who feel that crib-biting is directly linked to an increased risk of flatulent colic. This supports the need for good education programmes for horse owners when introducing new practises.

In some countries, stereotypic activities such as crib-biting and weaving are considered an 'unsoundness' when vetting horses (Hayes 1968), consequently must be declared at auction and tend to lower the value of affected animals (Mills pers comm). The reason for this is the popularly held view that other horses may mimic these and other stereotypies, creating additional problems for the owner. Because of the perceived risk of copying and because stereotypies are thought to be associated with health and performance problems, horses exhibiting them are often further isolated from other horses, for example stabled out of sight of conspecifics. This may exacerbate the behaviour if social isolation is a factor that increases stereotypy. Furthermore, isolating horses complicates management and contributes further to the unpopularity of stereotypic individuals and is a significant reason for their reduced market value.

In spite of the desire to prevent stereotypies for aesthetic and occasional health reasons, no traditional remedy that is effective for every crib-biter has yet been found. In the search for a permanent cure for stereotypic behaviour, all kinds of prevention are regularly tried. When the behaviour is established, the motivation to sterotype is so strong that the resourceful horse seems able to get around all devices aimed at thwarting it.

4.2. PREVENTION ATTEMPTS, INCLUDING SURGERY AND HARDWARE

4.2.1. Surgery

Attempts at long-term prevention of crib-biting by surgical intervention involve the excision of many combinations of different muscles and/or nerves of the ventral neck (Forsell 1929; Hamm 1977; Hakansson et al., 1992) but most have considerable drawbacks. With surgical approaches to crib-biting, especially myectomies, there is likely to be some disfigurement, which is particularly unwelcome in the case of show horses. Buccostomy wounds, on the other hand, regularly heal over well. Palatoschisis has also been described as a surgical approach to crib-biting (Smith 1924), but has fallen out of favour.

Differing success rates for myectomy are claimed by various authors, with Forsell (1929) quoting between 100 and 60 per cent success, while a more recent report (Hermans 1973) claims a 53 per cent 'cure' rate. Similar discrepancies appear in success rates cited for neurectomy, with Monin (1982) and Fraufelder (1981) claiming 60 per cent with successful outcomes and others reporting complete failure (Firth 1980). It seems likely that confusion in terminology, differing criteria of success, varying follow-up periods and post-operative management may all have contributed to this divergence in reports of success rates (Owen 1982; Schofield & Mulville 1998). Of the surgical cases that show partial improvement rather than complete resolution, grasping is reported to be more persistent than grunting. This supports the findings of McGreevy et al. (1995c), which revealed the involvement of the musculature of the ventral neck in the air-engulfing process that accompanies the characteristic grunt.

4.2.2. *Preventing the grasping of objects*
Short-term prevention of crib-biting can be accomplished by using bucket muzzles to eliminate the grasping component of the behaviour or by housing horses in boxes with no projections or modifying any ledges in a loose-box with rolling bars and plastic guttering (Kennedy et al., 1993). Anecdotal reports of cribbing horses in such projectionless boxes for three years (Hayes 1968) suggest that once acquired, the behaviour is extremely persistent. This reminds us that horses have good memories and their stereotypies are readily emancipated. The disadvantage of using a muzzle is that it may impair the ability of the horse to prehend food and interfere with diurnal ingestive rhythms. Similarly, devices that deliver dental and palatal pain may interfere with the horse's ingestive behaviour and result in a failure to thrive. While horses rapidly learn to eat with bits in their mouths, the presence of a permanent fistula is sometimes associated with transient drinking difficulties and the more persistent effusion of ingesta (Owen 1982).

4.2.3. *Interfering with air-engulfing*
Claims have been made that permanent fistulation of the buccal cavity (Karlander et al., 1965) or the use of a fluted bit will control the air-engulfing phase of both crib-biting and wind-sucking. These measures make it difficult for the horse to keep its mouth airtight. But since the mouth remains open during crib-biting, it would appear that consummation of the behaviour does not rely on a sealed buccal cavity. The reports claimed for the success of the buccostomy and fluted bit are therefore controversial. Furthermore, it has been suggested that oral structures are not critical to the air-engulfing phase of the behaviour.

4.2.4. *Punishing neck-flexion*
The most common means of short-term prevention of both crib-biting and wind-sucking is the application of a cribbing-collar (Hayes 1968). This simple device discourages the characteristic flexion of the horse's neck that accompanies the

behaviour. It consists of a leather strap incorporating a galvanised, hinged arc that accommodates the trachea and allows normal breathing despite the collar being tightened to the point where neck-flexing and/or the engulfing of air are not performed. Since the terminal grunting in this oral-based stereotypy is known to involve distension of the cranial oesophagus (McGreevy et al., 1995c), it is possible that this type of constriction not only makes crib-biting uncomfortable but also makes it less easy to consummate. Modifications of the collar include the use of leather spurs and metal spikes to increase the discomfort applied when the horse flexes its neck (Owen 1982). Often horses may adapt to the constriction of the collars, which are subsequently tightened, occasionally to the extent that skin trauma is apparent (Hachten 1995). Combinations of straps that can be tightened around different parts of the cranial neck have been advocated to reduce tissue damage at the poll (Hachten 1995).

4.2.5. *Punishing grasping*

Proprietary electric fencing is commonly used inside the stable as a deterrent to crib-biting. It is arranged on all ledges in the same way that taste deterrents are applied (Houpt & McDonnell 1993). There is an exhaustive list of taste deterrents for horses, ranging from sheepskin to creosote (Magner 1903; Miller & Robertson 1959). Meanwhile, filing of the incisor teeth, metal inserts between the teeth (Magner 1903) or others that impinge on the palate (Owen 1982) have been described as ways of making the grasping of fixed objects unpleasant.

Commercially available electronic dog-training collars have been adapted to fit the equine neck. These can be remotely controlled so that the horse does not associate punishment with the presence of a human (Houpt & McDonnell 1993). The principles of learning theory suggest that it would probably be advantageous to use dummy collars before and after aversion training of this sort in order to prevent the association of punishment with the collar itself. However, despite promising early reports, follow-up studies in 60 cases of aversion therapy using such electric shock collars indicated that only nine 'cures' were effected and of these three required reinforcement therapy after nine months (Owen 1982). One obstacle with the technique may be that the horse may not crib-bite for a long time after the first shock. Considerable patience and constant observation may be required for the trainer to maintain the contingency between crib-biting and punishment. It is reported that by applying an electric shock after the horse has grasped but before it has engulfed air, punishment and therefore extinction of grasping, rather than simple avoidance, may be achieved (Baker & Kear-Colwell 1974). The dose of shock is important, since learning may not occur above an optimal level of arousal (Lieberman 1993).

4.2.6. *Acupuncture*

While data for the success of acupuncture in the treatment of crib-biting are extremely limited, some success has been claimed for it in the treatment of crib-biters (Kuusaari 1983). Investigations into the use of ear staples to stimulate acupuncture points are proceeding (McDonnell, pers. comm.). It is important to acquire long term data to verify that short-term effects were not simply a result of the animal being distracted by the intervention.

4.2.7. *Pharmaceuticals*

Dopamine pathways are intimately related to the appearance of stereotypic behaviour, and opiate transmission regulates dopamine release (Cabib *et al.*, 1984). This has prompted research into the use of opioid antagonists as therapies for equine stereotypy, and sustainable pharmacological approaches to equine stereotypies are under investigation, for example, opioid antagonists (Dodman *et al.*, 1987). It has yet to be established, however, whether these remedies work by making the behaviours less rewarding or by reducing the frustrating effects of the environment. Transient elimination of crib-biting was reported in 100 per cent of subjects treated with naloxone, nalmefene or diprenorphine (Dodman *et al.*, 1987). Crib-biting was prevented for up to a week by continuous infusion of 5 to 10 mg of nalmefene per hour (Dodman *et al.*, 1987). But these pharmaceutical agents have many additional effects, for example, naloxone is a cataleptic that suppresses appetite, so they may also reduce the performance of other functional behavioural pathways.

The effects of such drugs on the horse's ability to perceive its environment are unclear. If they are used as a palliative to the effects of poor management, psychoactive pharmaceuticals may be contraindicated on welfare grounds (Houpt 1995, McGreevy & Nicol 1998a). The greatest impediment to the development of commercial forms of pharmaceutical cures appears to be the short half-life of their active ingredients (Dodman *et al.*, 1987). Some of the opioid antagonists would have to be injected daily, which would make them unacceptable to many owners. Subcutaneous depot preparations, still being developed, may be another option.

4.2.8. *Operant feeding*

Operant demand systems, using oral movements, for food rewards, have been investigated as an approach to the prevention of crib-biting (Houpt 1982; Winskill *et al.*, 1996; Henderson & Waran 2001). The premise is that this behaviour has its origin in frustration arising from low-arousal environments, so these devices can be used to deliver small aliquots of the horse's daily ration only when the horse has activated a trigger or moved a food-cube several times. Thus, the horse's time-budget can be replenished with behaviours that are neither stereotypic nor deleterious to its health. Data on the efficacy of operant feeding devices are limited but the approach has sufficient merit to be investigated further. One abiding problem may be that because operant demand devices rely on pelleted forms of food, they may fail to provide the level of oral stimulation that horses have evolved to require (Toates 1981; McGreevy *et al.*, 1995b).

4.2.9. *Turn-out to pasture*

It is widely acknowledged that the most successful means of reducing the frequency of all common stereotypies is to give the horse greater time in paddocks with *ad libitum* forage and social contact with conspecifics (Pell & McGreevy 1999). This option may, however, not be practical where there is limited access to grazing and may not completely erradicate the behaviour in all horses, as cribbing and weaving are also often seen in the field. Nevertheless, where possible this option would be the preferred means of treating stereotypies as it is effective at either eliminating or reducing stereotypy in virtually all cases by allowing horses to perform a wide variety of activities that may have been frustrated in the stable.

4.3. BEHAVIOURAL AND PHYSIOLOGICAL CONSEQUENCES OF PREVENTION

If stereotypies are a response to specific challenges faced in the stable environment then simply preventing the behavioural symptoms of the problem is no cure and this can result in a number of undesirable side effects, including perseverance despite the obstacles to performance, expression in a modified form and behavioural or physiological indicators of distress. For example, within an environment that severely limits normal forage intake (for example, an intensive training programme characterised by the provision of a high-concentrate : minimal-roughage diet), an oral stereotypy such as crib-biting may provide a route to normal feeding and digestive activity. Clearly, prevention of the symptom does nothing to ameliorate the cause. Horses prevented from crib-biting by the use of a traditional cribbing collar showed significantly more crib-biting on the first day after prevention than did control horses (McGreevy & Nicol 1998b). There was also a highly significant increase in the crib-biting rate of the test horses on the first day after prevention in comparison with their baseline rate. This defines the increase as a post-inhibitory rebound and it is argued that it reflects a rise in internal motivation to crib-bite during the period of prevention. Behaviours that exhibit this pattern of motivation are generally considered functional, and their prevention may compromise welfare.

It is possible that if a specific stereotypy is prevented, it may precipitate the appearance of unwelcome behavioural side-effects. Horses in electrified stables may, for example, exhibit greater reactivity, whilst horses in projectionless stables may be unable to adequately scratch or groom themselves and therefore their welfare may be further compromised. In addition, individual horses may perform more than one stereotypy and elimination of one stereotypy may precipitate the emergence of a modified or alternative stereotypy. For instance, horses without physical substrates for crib-biting occasionally develop the ability to crib-bite on their own limbs or the bodies of conspecifics (Hayes 1968; Boyd 1986). Alternatively, crib-biters prevented from grasping may begin to wind-suck (Sambraus & Rappold 1991). If crib-biting functions to reduce acidity of the digestive tract by the buffering action of equine saliva, as has been suggested by Nicol (1999b), then prevention of the opportunity to crib-bite may have harmful effects on the gut.

Finally, prevention of stereotypy per se may lead to distress, either because the activity is a general coping response to the captive environment or because prevention frustrates a highly motivated response to specific challenges encountered in the stable. Although evidence for the coping hypothesis in horses has been described as weak (Nicol 1999b), there is good evidence that preventing specific stereotypies causes physiological responses consistent with increased stress such as raised heart-rate and adreno-corticol activity. These elevated responses have been found both for physical prevention, for example weaving bars (McBride & Cuddeford 2001), and anti-cribbing collars (Lebelt *et al.*, 1998; McBride & Cuddeford 2001) and even the removal of favoured cribbing surfaces (McGreevy & Nicol 1998c). These responses, elevated heart-rate and corticosteroids and per-severance of the activity in forms that can further damage the horse, raise obvious concerns about the indiscriminate use of preventative measures. If stereotypies need to be reduced, then prevention alone is no solution, and treatment is more likely to be effective by focussing on the requirements of the stabled horse to resolve the underlying motivational problem or if this is not possible, to redirect the behav-iour to less harmful forms.

5. Preventative Management

5.1. SYMPATHETIC WEANING

Weaning can expose the young horse to a number of novel environmental chal-lenges, including removal from its dam, a change in nutritional environment, a change in its housing and changes in social environment. Horse breeders should do their utmost to reduce distress in foals at the time of weaning to avoid short or long term coping responses that may include overt redirected and stereotypic behaviours. Several studies have shown that group weaning has definite benefits in terms of incidence of stereotypy for the foals when compared with complete isolation (Houpt *et al.*, 1984; McCall *et al.*, 1985; Waters *et al.*, in press). The latter study in particular shows that both crib-biting and wood-chewing develop at an early age in foals, and that foals weaned individually are at a greater risk of devel-oping stereotypies than group-weaned foals. In terms of reducing the risks of stereo-typy, paddock weaning is therefore recommended and stable weaning in isolation should be avoided. Even in group weaned foals, careful attention should be paid to post weaning feed. Foals receiving a grain-based hard-feed following weaning were four times as likely to develop cribbing as foals who did not receive a hard feed (Waters *et al.*, in press). Finally, even the type of forage can be a risk factor, as those receiving hay as post-weaning forage, were less likely to develop wood-chewing than those on a hay replacer such as haylage (Waters *et al.*, in press).

5.2. ENVIRONMENTAL ENRICHMENT

While the concept of emancipation explains why it may not be possible to reverse the process of stereotypy development in mature horses, the proportion of time they spend performing the stereotypy can be reduced by increasing opportunities for foraging behaviour (Houpt & McDonnell 1993) and social contact (Cooper *et al.*, 2000). Remedial steps to improve the environment in which an animal performing stereotypic behaviour is in, should be encouraged, simply because this has the potential for improving its welfare.

Given the social structure and time budgeting of feral horses, the importance of the roles played by social contacts and foraging could have been predicted. Stereotypy prevalence were lower in Australian Thoroughbred horses kept at pasture compared with stabled horses (Pell & McGreevy 1999). This statistical information is important because it suggests that some stereotypic behaviours may be caused by management practices. Management factors provide useful indices of what may be sub-optimal in the environment of intensively housed performance horses. These revolve around the stable management of young animals and how much of their day is spent experiencing this environment. Stable designs that enhance the degree of social contact between horses, the use of straw bedding, and a generous daily forage ration that is preferably placed in a haynet with fine mesh are all considered likely to reduce the incidence of new stereotypies and to lessen the degree to which established stereotypies are performed.

Horses choose to perform much of their weaving over the stable door, especially when companions from neighbouring stables are removed. This suggests that horses may start to weave when their motivation to leave the enclosure with other members of the herd is thwarted. Stable designs that reduce social isolation are associated with a risk of stereotypic behaviour (including weaving), so there is support for the view that weaving develops as a form of frustrated escape response (Kiley-Worthington 1987). Furthermore, weaving was significantly reduced when weaving horses were housed in stables that provided increased visual contact with neighbouring horses. The incidence of weaving dropped to zero when the horses had opportunities for social interaction with their neighbours on all four sides of their enclosure (Cooper et al., 2000).

5.3. DIETARY MANAGEMENT, INCLUDING ADDITIVES

It has been known for some time that the feeding of concentrate diets (Rowe et al., 1994) and periods of food deprivation increase gastric acidity to harmful levels that can result in rapid ulceration (Murray & Eichorn 1996). High-concentrate diets also alter caecal fermentation and increase caecal acidity (Willard et al., 1977). Early signs of abnormal wood-chewing and stable biting were higher in horses fed a predominantly concentrate diet than horses fed 8 kg of hay per day. However, this did not remain true if the horses fed concentrates were given an additional supplement, virginiamycin, which suppresses lactic acid production in the hindgut and increases hindgut pH (Johnson et al., 1998). This suggests that crib-biting may function to reduce acidity of the digestive tract. The exact mechanism by which crib-biting could reduce the acidity of the digestive tract is not known but it has been suggested that this activity may result in increased salivary flow (Nicol 1999b).

5.4. BREEDING STRATEGIES AND DATABASES

Since we know that some unwelcome behaviours follow a familial pattern of inheritance (Vecchiotti & Galanti 1986) it may be that there will emerge a drive to breed from animals that can cope with the stressors of intensive management. While there may be initiatives to identify the genes of horses that cope best with intensive management and are thus less likely to display stereotypies, it is hoped that the technology can be harnessed to improve horse welfare rather than simply to select for tolerance.

The prevalence of stereotypies puts them on a par with other disease processes, for example, lameness, which in 1982 was found to end 10.6 per cent of racing careers (Jeffcott et al., 1982). However, because they are not a direct cause of wastage and perhaps because they do not alter the appearance of the animal, they prompt few attempts at preventive medicine. The increasingly prevalent practice of identifying and recording the performance of non-racing sport horses, for example, the British Horse Database, will enhance the quality and, hopefully, the depth of the gene pool. The results of the surveys may provide methods for change, while the incentives for change may remain more obscure. The economics of performance animal welfare may be more difficult to calculate than those of farm animals (Bennett 1995; McInerney 1995).

6. Conclusion

Epidemiological and empirical studies consistently indicate an association between equine stereotypies and two management factors in particular, the feeding of low fibre/high grain content feeds and restrictions on social behaviour. In conventional equine husbandry, both differ markedly from the ecological niche for which the horse has evolved and both suggest that stereotypy is a response to behavioural restriction in the stable. There is also evidence that stereotypic activities can be conditioned responses to rewarding situations in the stable and that early experience can have a significant effect on both the form and likelihood of stereotypy.

To prevent the performance of equine stereotypies, humans have used physically restricting devices, electric shock and surgery. In many cases these methods are distressing, harmful and ineffective, especially where they address the symptoms rather than the underlying cause. Reduction of stereotypic behaviour by allowing appropriate expression of the underlying motivations is a better strategy than control. By feeding high forage rations, providing the minimum amount of concentrate feed to sustain health, and maximising opportunities for social contact, especially around the time of weaning, the emergence of new stereotypies and expression of established ones should be reduced.

7. References

Baker, G.J. and Kear-Colwell, J. (1974) Aerophagia (wind-sucking) and aversion therapy in the horse. *Proceedings of the American Association of Equine Practitioners* **20**, 127–130.

Bennett, R. (1995) Estimating the perceived benefits of measures to improve animal welfare, in E.A. Goodall (ed.), *Proceedings of the Society for Veterinary Epidemiology and Preventive Medicine.* University of Reading, Reading, UK.

Broom, D.M. and Johnson, K.G. (1983) *Stress and Animal Welfare.* Chapman and Hall, London, UK.

Cabib, S., Puglisi-Allegra, S. and Oliveria, A. (1984) Chronic stress enhances apomorphine-induced stereotyped behavior in mice: involvement of endogenous opioids. *Brain Research* **298**, 138–140.

Cooper, J.J. and Nicol, C.J. (1994) Neighbour effects on the development of stereotypic behaviour in bank voles (*Clethrionymus glareolus*). *Animal Behaviour* **47**, 214–216.

Cooper, J.J., Odberg, F.O. and Nicol, C.J. (1996) Limitations of the effect of environmental improvement in reducing stereotypic behaviour in bank voles (*Clethrionomys glareolus*). *Applied Animal Behaviour Science* **48**, 237–248.

Cooper, J.J. and Mason, G.J. (1998) The identification of abnormal behaviour and behavioural problems in stabled horses and their relationship to horse welfare: a comparative review. *Equine Veterinary J. Suppl.* **27**, 5–9.

Cooper, J.J., McDonald, L. and Mills, D.S. (2000) The effect of increasing visual horizons on stereotypic weaving: implications for the social housing of stabled horses. *Applied Animal Behaviour Science* **69**, 67–83.

Dodman, N.H., Shuster, L., Court, M.H. and Dixon, R. (1987) Investigation into the use if narcotic antagonists in the treatment of a stereotypic behaviour pattern (crib-biting) in the horse. *American J. Veterinary Research* **48**, 311–319.

Firth, E.C. (1980) Bilateral ventral accessory neurectomy in windsucking horses. *Veterinary Record* **106**, 30–32.

Forsell, G. (1929) Kopen, chirurgische Behandlung. *Stang-Wirth, Tierheil-kunde und Tierzucht* **6**, 283–287.

Fraufelder, H. (1981) Treatment of crib-biting: A surgical approach in the standing horse. *Equine Veterinary J.* **13**, 62–63.

Green, P. and Tong, J.M.J. (1988) Small intestinal obstruction associated with wood chewing in two horses. *Veterinary Record* **123**, 196–198.

Hachten, W. (1995) Cribbing treatment. *The Equine Athlete* **8**, 20–21.

Hakansson, A., Franzen, P. and Petersson, H. (1992) Comparison of two surgical methods for treatment of crib-biting in horses. *Equine Veterinary J.* **24**, 494–496.

Hamm, D. (1977) A new surgical procedure to control crib-biting. *Proceedings of the American Association of Equine Practitioners* **23**, 301–302.

Harris, P.A. (1999) How understanding the digestive process can help minimise digestive disturbances due to diet and feeding practices. In Harris, P.A., Gomarsall, G.M., Davidson, H.P.B. and Green, R.E. (eds.), *Proceedings of the British Equine Veterinary Association Specialist Days on Behaviour and Nutrition*, pp. 45–49.

Hayes, M.H. (1968) *Veterinary Notes for Horse-owners*, Sixteenth Edition. Stanley Paul, London.

Henderson, J.V. and Waran, N.K. (2001) Reducing equine stereotypies using the Equiball™. *Animal Welfare* **10**, 73–80.

Hermans, W.A. (1973) Kribbebijten-luchtzuigen. *Tijdschrift voor Diergeneeskunde* **22**, 1132–1137.

Houpt, K.A. (1982) Oral vices of horses. *Equine Practice* **4**, 16–25.

Houpt, K.A. (1995) New perspectives on equine stereotypic behaviour. *Equine Veterinary J.* **27**, 82–83.

Houpt, K.A., Hintz, H.F. and Butler, W.R. (1984) A preliminary study of two methods of weaning foals. *Applied Animal Behaviour Science* **12**, 177–1881.

Houpt, K.A. and McDonnell, S.M. (1993) Equine Stereotypies. *The Compendium of Continuing Education for the Practising Veterinarian* **15**, 1265–1272.

Hughes, B.O. and Duncan, I.J.H. (1988) The notion of ethological 'need', models of motivation and animal welfare. *Animal Behaviour* **36**, 1696–1707.

Jeffcott, L.B., Rossdale, P.D., Freestone, J., Frank, C.J. and Towers-Clark P.F. (1982) An assessment of wastage in Thoroughbred racing from conception to 4 years of age. *Equine Veterinary J.* **14**, 185–198.

Johnson, K.G., Tyrrell, J., Rowe, J.B. and Petherick, D.W. (1998) Behavioural changes in stabled horses given non-therapeutic levels of virginiamycin. *Equine Veterinary J.* **30**, 139–142.

Karlander, S., Mansson, J. and Tufvesson, G. (1965) Buccostomy as a method of treatment for aerophagia (wind-sucking) in the horse. *Nordisk Veterinär Medicin* **17**, 455–458.

Kennedy, J.S., Schwabe, A.E. and Broom, D.M. (1993) Crib-biting and wind-sucking stereotypies in the horse. *Equine Veterinary Education* **5**, 142–147.

Kiley-Worthington, M. (1983) Stereotypies in the horse. *Equine Practice* **5**, 34–40.

Kiley-Worthington, M. (1987) *The Behaviour of Horses in Relation to Management and Training.* J.A. Allen, London, UK.

Krzak, W.E., Gonyou, H.W. and Lawrence, L.M. (1991) Wood chewing by stabled horses: diurnal pattern and effects of exercise. *J. Animal Science* **69**, 1053–1058.

Kuusaari, J. (1983) Acupuncture treatment of aerophagia in horses. *American J. Acupuncture* **11**, 363–370.

Lebelt, D., Zanella, A.J. and Unshelm, J. (1998) Physiological correlates associated with cribbing behaviour in horses; changes in thermal threshold, heart rate, plasma B-endorphin and serotonin. *Equine Veterinary J. Suppl.* **27**, 21–27.

Lieberman, D.A. (1993) *Learning: Behaviour and Cognition.* Brooks/Cole Publishing Company, Pacific Grove, California, UK.

Luescher, U.A., McKeown, D.B. and Dean, H. (1998) A cross-sectional study on compulsive behaviour (stable vices) in horses. *Equine Veterinary J. Suppl.* **27**, 14–18.

McAfee, L.M, Mills, D.S. and Cooper, J.J. (in press) The use of mirrors for the control of stereotypic weaving behaviour in the stabled horse. *Applied Animal Behaviour Science.*

McBride, S.D. and Cuddeford, D. (2001) The putative welfare-reducing effects of preventing equine stereotypic behaviour. *Animal Welfare* **10**, 173–189.

McBride, S.D and Long, L. (2001) Management of horses showing stereotypic behaviour, owner perception and the implications for welfare. *Veterinary Record* **148**, 799–802.

McCall, C.A., Potter, G.D. and Kreider, J.L. (1985) Locomotor, vocal and other behavioural responses to varying methods of weaning foals. *Applied Animal Behaviour Science* **14**, 27–36.

McFarland, D.J. (1989) *Problems of Animal Behaviour.* Longman, Harlow, UK.

McGreevy, P.D. (1995) *The Functional Significance of Stereotypies in the Stabled Horse.* PhD thesis. Department of Animal Husbandry, University of Bristol.

McGreevy, P.D. (1996) *Why Does My Horse . . . ?* Souvenir Press, London, UK.

McGreevy, P.D., Cripps, P.J., French, N.P., Green, L.E. and Nicol, C.J. (1995a) Management factors associated with stereotypic and redirected behaviour in the Thoroughbred horse. *Equine Veterinary J.* **27**, 86–91.

McGreevy, P.D., Richardson, J.D., Nicol, C.J. and Lane, J.G. (1995b) A radiographic and endoscopic study of horses performing an oral stereotypy. *Equine Veterinary J.* **27**, 92–95.

McGreevy, P.D., French, N.P. and Nicol, C.J. (1995c) The prevalence of abnormal behaviours in dressage, eventing and endurance horses in relation to stabling. *Veterinary Record* **137**, 36–37.

McGreevy, P.D. and Nicol, C.J. (1998a) Prevention of crib-biting: a review. *Equine Veterinary J. Suppl.* **27**, 35–38.

McGreevy, P.D. and Nicol, C.J. (1998b) Physiological and behavioural consequences associated with short-term prevention of crib-biting in horses. *Physiology and Behaviour* **65**, 15–23.

McGreevy, P.D. and Nicol, C.J. (1998c) The effect of short term prevention on the subsequent rate of crib-biting in Thoroughbred horses. *Equine Veterinary J. Suppl.* **27**, 30–34.

McGreevy, P.D., Webster, A.J.F. and C.J. Nicol. (2001) A study of the digestive efficiency, behaviour and gut transit times of crib-biting horses. *Veterinary Record* **148**, 592–596.

McInerney, J.P. (1995) The cost of welfare. In Goodall, R.E. (ed.), *Proceedings of the Society for Veterinary Epidemiology and Preventive Medicine.* University of Reading, Reading, UK.

Magner, D. (1903) *Magner's Standard Horse and Stock Book.* Saalfield, New York.

Marsden, M.D. (1993) Feeding practices have greater effect than housing practices on the behaviour and welfare of the horse. In Collins, E. and Boon, C. (eds.), *Proceedings of the 4th International Symposium on Livestock Environment.* American Society of Agricultural Engineers, University of Warwick, Coventry, pp. 314–318.

Mason, G.J. (1991) Stereotypies: a critical review. *Animal Behaviour* **41**, 1015–1037.

Mason, G.J. and Mendl, M. (1997) Do the stereotypies of pigs, chickens and mink reflect adaptive species differences in the control of foraging? *Applied Animal Behaviour Science* **53**, 45–58.

Miller, W.C. and Robertson, E.D.S. (1959) *Practical Animal Husbandry,* Seventh edition. Oliver and Boyd, Edinburgh.

Mills, D.S. (1998) Personality and individual differences in the horse. their significance, use and measurement. *Equine Veterinary J. Suppl.* **27**, 10–13.

Mills, D. and Nankervis, K. (1999) *Equine Behaviour: Principles and Practice.* Blackwell Science, Oxford.

Mills, D.S, Eckley, S. and Cooper, J.J. (2000) Thoroughbred bedding preferences, associated behaviour differences and their implications for equine welfare. *Animal Science* **70**, 95–106.

Mills, D.S. and Davenport, K. (in press) The effect of a neighbouring conspecific versus the use of a mirror for the control of stereotypic weaving behaviour in the stabled horse. *Animal Science.*

Monin, T. (1982) Surgical management of crib-biting in the horse. *Compendium on Continuing Education of the Practicing Veterinarian* **4**, 69–72.

Murray, M.J. and Eichorn, E.S. (1996) Effects of intermittent feed deprivation, intermittent feed deprivation with ranitidine administration, and stall confinement with ad libitum access to hay on gastric ulceration in horses. *American J. Veterinary Research* **11**, 1599–1603.

Nicol, C.J. (1999a) Stereotypies and their relation to management. In Harris, P.A., Gomarsall, G.M., Davidson, H.P.B. and Green, R.E. (eds.), *Proceedings of the British Equine Veteinary Association Specialist Days on Behaviour and Nutrition,* pp. 11–14.

Nicol, C.J. (1999b) Understanding equine stereotypies. *Equine Veterinary J. Suppl.* **28**, 20–25.

Nicol, C.J., Wilson, A.D., Waters, A.J., Harris, P.A. and Davidson, H.P.B. (2001) Crib-biting on foals in associated with gastric ulceration and mucosal imflammation. In Garner, J.P., Mench, J.A. and Heekin, S.P. (eds.), *Proceedings 35th International Society for Applied Ethology Congress.* Centre for Animal Welfare; Califronia, USA.

Owen, R.R. (1982) Crib-biting and wind-sucking – that equine enigma. In Hill, C.S.G. and Grunsell, F.W.G. (eds.), *The Veterinary Annual 1982.* Wright Scientific Publications, Bristol, UK, pp. 156–168.

Pell, S. and McGreevy, P.D. (1999) The prevalence of abnormal and stereotypic behaviour in Thoroughbreds in Australia. *Australian Veterinary J.* **77**, 678–679.

Powell, S.B., Newman, H.A., McDonald, T.A., Bugenhagen, P. and Lewis, M.H. (2000) Development of spontaneous stereotyped behavior in deer mice: Effects of early and late exposure to a more complex environment. *Developmental Psychobiology* **37**, 100–108.

Redbo, I., Redbo-Torstensson, P., Odberg, F.O., Hedendahl, A. and Holm, J. (1998) Factors effecting behavioural disturbances in race-horses. *Animal Science* **66**, 475–481.

Ritzberger-Matter, G. and Kaegi, B. (1998) Retrospective analysis of the success rate of surgical treatment of aerophagia in horse at the Veterinary Surgical Clinic, University of Zurich. *Equine Veterinary J. Suppl.* **27**, 62.

Rowe, J.B., Pethick, D.W. and Lees, M.J. (1994) Prevention of acidosis and laminitis associated with grain feeding in horses. *J. Nutrition* **124**, 2742–2744.

Sambraus, H.H. and Rappold, D. (1991) Crib-biting and wind-sucking in horses. *Pferdeheilkunde* **7**, 211–216.

Schofield, W.L. and Mulville, J.P. (1998) Assessment of the modified Forsell's procedure for the treatment of oral stereotypies in 10 horses. *Veterinary Record* **142**, 572–575.

Smith, F. (1924) An attempt to deal with 'windsucking' by a surgical interference. *Veterinary J.* **80**, 238-240.

Toates, F.M. (1981) The control of ingestive behaviour by internal and external stimuli – A theoretical review. *Appetite* **2**, 35–50.

Vecchiotti, G. and Galanti, R. (1986) Evidence of heredity of cribbing, weaving and stall-walking in Thoroughbred horses. *Livestock Production Science* **14**, 91–95.

Waring, G.H. (1983) *Horse Behaviour. The Behavioural Traits and Adaptations of Domestic and Wild Horses, Including Ponies.* Noyes Publications New York, USA.

Waters, A.J. (2002) *The Development of Stereotypic Behaviour in Thoroughbred Horses.* PhD thesis. University of Bristol, UK.

Waters, A.J., Nicol, C.J. and French, N.P. (in press) The development of stereotypic and redirected behaviours in young horses: the findings of a four year prospective epidemiological study, *Equine Veterinary J.*

Wiepkema, P.R. (1985) Abnormal behaviours in farm animals: ethological implications. *Netherlands J. Zoology* **35**, 279–299.

Willard, J.G., Willard, J.C., Wolfram, S.A. and Baker, J.P. (1977) Effect of diet on cecal pH and feeding behaviour of horses. *J. Animal Science* **45**, 87–93.

Winskill, L.C., Young, R.J., Channing, C.E., Hurley, J. and Waran, N.K. (1996) The effect of a foraging device (a modified Edinburgh foodball) on the behaviour of the stabled horse. *Applied Animal Behaviour Science* **48**, 25–35.

Chapter 6

THE EFFECTS OF TRANSPORTATION ON THE WELFARE OF HORSES

N. WARAN
Department of Veterinary Clinical Studies, Royal (Dick) School of Veterinary Studies, University of Edinburgh, Easter Bush, Roslin, Midlothian EH25 9RG, UK

D. LEADON
Irish Equine Centre, Johnstown, Naas, Co. Kildare, Ireland

T. FRIEND
Department of Animal Science, Texas A&M University, College Station 77843, Texas, USA

Abstract. Typically, horses are transported many times in their lives, this is with the exception of the horses reared for meat. Although difficult to estimate the extent of the movement of horses world-wide, it is clear that this is a substantial and growing practice. Until recently research into the effects of the different methods of transport (road, sea and air), was limited. This may have been because it was presumed that, because of their financial and emotional value, horses experience higher standards of transportation, than other large domestic animals. The process of transporting horses includes a range of potential stressors, and there is scientific evidence that many of these can impact upon the welfare of the horse. In this chapter, we examine the effects of the different modes used to transport horses and we offer suggestions where possible for improvements in this practice.

1. Introduction

Horses are transported for a variety of reasons including sport/recreation, breeding, meat production and slaughter. Apart from those horses reared for meat, most horses are transported several times in their lives. This differs from most other large domestic animals. It is difficult to determine the actual numbers of horses being transported in the various countries. In the UK there are an estimated 565,000 'riding/sports' horses (BHS 1998), and in the US there are an estimated 5.2 million horses. In 1992, 318,509 horses were moved across EC national frontiers. This number is likely to have increased substantially in recent years, and only reflects movement across national boundaries and not within each country. When it comes to commercial transport of horses for sale and competition this is also difficult to estimate, but one UK based business estimated that they transported 50 horses per week between Scotland and the South of England (Gillie, pers. comm. 1996). Horse owners in Texas transport an average of 2.5 horses (Gibbs *et al.*, 1997) on 24 trips per year (Gibbs *et al.*, 1998). There are many commercial transporters within each country but since they are for the most part unregulated, it is impossible to determine the size of the industry.

N. Waran (ed.), The Welfare of Horses, 125–150.

Horses have been and still are transported by land (road and rail), sea and air. Each of these modes of transport will impact upon the horse in a different way. In this chapter we examine what is known about the way in which the different methods used for transporting horses affect their welfare and discuss various methods for improving the situation.

The transportation of horses has a long history. It is claimed that as early as 480 B.C. Xerxes moved his entire cavalry by ship across the Hellepont against the Ancient Greeks. Hannibal is recorded as having used rafts to get his horses across the Rhone in 218 B.C. and later in the 14th Century AD, a two-way trade in horses grew between Britain and the Continent of Europe. For the most part the reasons for transporting horses were linked with their use by the military.

Early forms of transportation for horses were mostly water-borne, but land based methods for moving horses were developed from the 18th Century onwards. Following World War I, reliance upon horsepower reduced dramatically in most of the industrialised countries. In addition motor vehicles had begun to replace the use of the horse in urban areas as early as 1910 (Barker 1983). The change to an animal used increasingly for recreation, rather than as a traction or military animal, led to a change in the modes of transport used for horses. Transport by road is now commonly used for horses, but increasingly horses are being moved across and between countries by air.

2. Transport by Rail

Before the invention of the internal combustion engine, land-based horse transport was either oxen/horse powered or by rail. In the United Kingdom, rail systems were developed in the 1840's and most stations had loading ramps for use by stock being moved to markets. Concern for the welfare of animals during rail transport became apparent, probably due to the injuries incurred and the financial consequences, during 1895. The Animals (Transit and General Order) (HMSO 1895) in the UK listed certain compulsory design features to aid the animals' balance. Buffers, bedding and battens (foot holds) were mentioned but no mention was made of the various ways used to tie or restrain the animals. Ventilation and protection from the weather were mentioned in the regulations issued by the Board of Agriculture in 1904, and construction specifications for vehicles used in equine transport were laid down. In 1913, 135,265 horses were transported by rail in the UK (Board of Trade 1915), with charges for carriage being based on the amount of floor space occupied and type of horse. Race horses were transported in special 'Race horse vans' which were added to passenger trains, but other types of horse were carried in 'horseboxes'. The decline in the number of horses transported by rail in the UK due to the improved road network, led in 1969 to all facilities for the loading and carriage of horses by rail being withdrawn. However in Europe, transport by rail still occurs (Larter & Jackson 1987). In North America, the only horses regularly transported by rail accompany the two Ringling Brothers and Barnum & Bailey circuses. Since transport by rail is likely to be used less frequently, as road transport systems improve both within and between countries it will not be considered further. However where horses are still transported by rail the sections that follow on environmental conditions, restraint and confinement and food/water and rest will all apply.

3. Road Transport

3.1. LEGISLATION

As AATA (2000) points out, legislation is updated and amended constantly. The basis for all of the transport legislation within Europe is the European convention for the protection of animals during international transport (signed in Europe by 8 countries in 1968). At this moment there are 23 countries signed up and a further 19 countries that may sign the convention soon. Countries within the European Union are expected to prepare national legislation using the appropriate European Directive as the minimum standard. The European Directive 95/29/EC currently outlines the minimum requirements for the transport of horses. The specific details relating to horses are contained within Recommendation no. R(87)17, on the transport of horses (September 1987). Loading densities, container requirements, feed and water requirements, travelling times and staging points are all described. These are very clearly outlined in Chapter 8.6 of the AATA Manual for the transportation of live animals (Harris 2000). As an example, in the UK, 'The welfare of animals (Transport) Order 1997' (WATO), (HMSO 1997) interprets the European directive, and puts into place national requirements for the transport of horses within the boundaries of the UK.

Other countries outside the EU will have their own legislation. In Australia, the transport of horses is controlled through the 'Land transport of horses 7', which is an Australian Model Code of Practice. States within Australia are expected to use this to design their own codes of practice that meet individual needs. Although the US has one of the first federal laws regulating the transport of livestock by rail and boat, the 1906 '28-Hour Law', this law has not been amended to cover transport by truck. The 'Safe Commercial Transportation of Equine to Slaughter Act' that was passed as part of the 1996 US Farm Bill authorised the USDA (United States Department of Agriculture) to establish regulations for the transport of slaughter horses. Although the USDA drafted regulations based partly on a series of studies (reviewed by Friend 2001) funded by the USDA, the final regulations have been delayed by animal activists who believed that the regulations were too limited, and by the change to a more conservative administration in 2001. States are free to enact their own regulations as long as they meet or exceed federal regulations. In New York, for example, the transport of horses in double-decked trailers has been banned.

A major problem is the enforcing of transport codes/regulation. There are no longer border controls between EU countries, which makes it possible to drive for extremely long distances in adverse weather conditions, without stopping. Staging points within the EU are approved premises where animals can be unloaded, fed, watered and rested. Council Regulation (EC) No. 1255/97 contains the specific requirements, such as health and hygiene measures, facilities and operations. Transport operators are expected to draw up transport schedules, including routes taken, and travelling and resting times. However, the use of the staging points and the accuracy of the transport records must be checked. Despite this, the EU controls on the transport of animals are probably the most advanced in the world in protecting animal welfare.

3.2. PROBLEMS WITH ROAD TRANSPORT

Until recently, the transport of horses by road had received limited scientific investigation. Horses are often thought to enjoy a higher standard of transportation than other farmed animals, but this is not always the case (Waran 1993). Horses used for sport or kept as companion animals (pets) may well receive special care during transport but horses transported for slaughter are likely to experience similar conditions to the other domestic farmed species. The potential stressors associated with road transport include loading, unloading, confinement, isolation/inappropriate grouping, and the motion of the vehicle, environmental conditions, amongst others. These are the factors that also affect other farmed species. Several studies have indicated that transportation by road can be associated with various behavioural, physiological and immunological responses. Due to the complexity of the transportation process it is often difficult to assess which component of the experience has caused the response.

Anecdotal evidence suggests that the transport process can be detrimental to welfare. Cregier (1981), reported observations in which transport appeared to be associated with the incidence of acute colitis, laminitis, transit tetany, trailer choke and mild azoturia, but the linkage with transport has yet to be confirmed by scientific study.

Transport by road can be associated with an increase in plasma cortisol (Friend *et al.*, 1998) and progesterone (Baucus *et al.*, 1990; von Mayr and Siebert 1990; White *et al.*, 1991). Increased sodium levels and blood glucose concentrations have also been measured in slaughter horses (Friend *et al.*, 1998; Friend 2000) and experienced racehorses (White *et al.*, 1991). Elevated heart rates (Smith 1992; Waran 1993) and increased concentrations of AST and CK (the two muscle enzymes, Aspartate amino transferase and Creatine kinase respectively) have also been reported (see Leadon *et al.*, 1989). In addition, various conditions have been shown to be provoked by the transportation process, amongst these are hyperlipaemia (Forhead *et al.*, 1990), reactivation of salmonella infection (Owen *et al.*, 1983) as well as various respiratory problems (Leadon *et al.*, 1990). It is true that the physical condition that the horse is in prior to transport will influence the ability to cope with any stressors. Horses transported for slaughter are likely to be in poor health/condition and as such may be expected to be more susceptible (Friend *et al.*, 1998). In addition experience will influence the extent to which a horse is affected by being transported. Horses used for sporting/recreational purposes that

have had a number of relatively positive experiences of being loaded and transported are less likely to be adversely affected than those with no experience and
those who have had a negative previous experience (such as a fall, over-crowding
etc.). However, those horses that may have become habituated to transportation by
road, may also suffer from problems associated with frequent journeys. The pressures placed upon the frequent traveller are likely to be associated with the physical
demands associated with fatigue, disrupted feeding patterns, loss of weight,
restricted movement and so on. Despite this, frequent transport may provide positive
stimulation. For example, ranch horses, rodeo horses, circus horses and performance
horses are transported many times in their lives with few problems.

3.3. METHODS OF ROAD TRANSPORT

Not only are there various reasons for transport and along with this different
standards, but there are also different types of road vehicles. The motorised horse
box or horse lorry can vary in capacity from one to more than 10 individually stalled
or grouped horses. These lorries (wagons, vans or trucks) range from closed wooden
or metal-sided lorries, to open-topped and even double-decked vehicles. The
standard of design and maintenance of the ramps, floors, ventilation system and
internal furniture varies from the basic to the luxurious. An alternative mode of
road transport is the horse trailer (boxes or floats). This is a popular method for
transporting sports/recreation horses since they are usually cheaper than the cost
of buying and maintaining a dedicated motorised vehicle. There are different types
of trailer depending on use; the 'stock' trailer will carry a large number of loose
(i.e. not tied or individually stalled) horses, whereas the competition trailer is usually
designed for two or three horses. The trailer has a rear door/ramp and sometimes
a side ramp. The horses face towards or sideways to the direction of travel, and
the container is towed by a car or van. Trailers appear to have some disadvantages; they are towed behind a vehicle, and so it is difficult to check conditions
inside the container, and there is little room in them in times of an emergency. They
are usually attached to the towing vehicle at one point via a tow hitch, (ball socket
or bar), which makes them particularly susceptible to the vectoral changes acting
across the hitch when on minor roads with many bends/corners. There are other
designs where the attachment point sits forward of the rear axle of the towing
vehicle. These 'goose-neck' or tractor-type trailers are considered to be more stable
for towing than the ball-hitch, bumper-pull types. It is however probable that horses
transported in all types of trailers are more susceptible to driver ability than those
in lorries. The trailer is likely to 'sway' from side to side more, and there is a risk
of it being over-turned. Despite this, the well built, double horse trailer driven by
a competent person is often used as a horse ambulance, as it has low slung axles
which means that the loading ramp is less steep than a lorry/truck. In addition any
vehicle can tow the trailer in an emergency, and various internal modifications
(slings, padding) and external features (winch, moving floors) can be easily added.
Although trailers are now designed so that they can carry more than two animals
in various different orientations (rear-face, forward-face and herring-bone/slant

(sideways), it is still common practice within the UK and other parts of Europe to carry two horses facing the direction of travel. However, aluminium slant load trailers with a capacity of two to four horses are now the most popular in the US and Canada. This has implications for a horse's ability to balance, ease of loading and air quality (see later sections in this chapter). Double-decked horse trailers are commonly used in the US and Canada, although their use may be phased out. Double-decked horse trailers are modified so that there is more head clearance in the compartments than cattle trailers, but there is still not adequate height for larger horses and these types of trailers are associated with increased injuries (Stull 1999; Grandin 2000).

4. Components of Transport

There is considerable variation amongst horses in their response to the transportation process and in studies of the effects of transport by road on farm species, the process has been divided into various components (see Stephens *et al.*, 1985, Philips *et al.*, 1988). Transportation consists of; loading, restraint and confinement in the vehicle, movement and vibration of the vehicle, internal environment (design factors and physical environment), effects of food, water and rest opportunities, unloading and recovery (see Waran 1993).

4.1. LOADING

Loading is considered to have a great impact on animals to be transported (see Waran 1993). Studies of other species (*e.g.* pigs, sheep) have demonstrated that heart rate increases are proportional to the angle of the ramp (Warriss *et al.*, 1991) and steeper ramps are avoided (Philips *et al.*, 1991). Baldock and Sibly (1990) point out that although a rise in heart rate can be attributed to the effort required to climb the ramp, there are also substantial non-motor increases associated with the loading process. In horses it has been suggested that fear of entering into an enclosed space as well as the height of the step up onto the ramp, and the instability and incline of the ramp, lead to fear of loading (Houpt 1982). Ramp climbing was investigated by Waran and Cuddeford (1995). In this study, horses under 3 years were less willing to load (yearlings took a median of 368 seconds to load as compared with the over three year olds at 5 seconds), showed more evasive behaviour (such as plants (refusing to move) swings (moving the body to the side) and pull-backs (reversing)) and had higher peak heart-rates than older more experienced animals. The step-up into the vehicle appeared to cause problems for the young horses, and before committing to loading, horses would nose and even paw at the ramp. This behaviour was observed and commented upon by Hayes as early as 1902, when he noticed that equines being loaded via gang-planks onto ships would sniff the gang-plank before loading. Horses tend to be neophobic, and as such it is not surprising that they will tend to avoid placing themselves in a novel situation in which they are unable to escape easily. Interestingly even when a loading platform is used so that the horse does not have to climb, heart rates are still high at loading.

Houpt (1982) suggests that all horses should be accustomed to loading as foals. Reducing the step up onto the ramp and the ramp incline will encourage ease of loading. It has been suggested that reversing the horse into a trailer (via a flat ramp) induces fewer fear responses. Making the ramp more solid, so that it appears more stable may also help (Bush 1992), and rubber matting may buffer the hollow sound of a ramp when it is being used. Hydraulic decks or lifts are used for some farm species (Harris 1996), but not for horses. It seems unlikely that they will become popular due to the expense, and the problems with the likely response of a horse to standing on a vertically moving platform. It is common practice for trailers in the US and Canada to have gates or doors to the rear of the vehicle instead of a ramp. The doors open outwards so that the horse steps up into the trailer. Apart from the reduced cost of such means of entry, a further advantage of the step-up trailer design is thought to be that it ensures that the horse lowers its head upon entry to the container. This protects the horse against potential damage to its poll/head, and may also reduce the horse's fear of loading. In order to ensure that young or poorly trained horses are guided into the vehicle, it is usual for there to be side gates or doors attached to either side of the ramp. In order to avoid damage to the horse the gap or space between the ramp and the bottom side of these side gates should be large enough so that if a horse does slip off the ramp it does not become trapped. Alternatively the use of 'stock-gates', that can be lowered and attached to the ramp sides, reduces the chances of a horse becoming trapped, and will funnel the horse's entry to the vehicle more effectively. Finally there is the issue of lighting. Many animals have been shown to be more willing to enter a well lit area (going from dark to light) than a dark cavern. Although the evidence for this is limited, and in fact no such evidence was found whilst loading pigs (Philips et al., 1988), there appears to be a good foundation for this claim and anecdotal evidence of its effectiveness (Houpt 1982). In addition to attempting to make the vehicle and loading aspects more attractive to a horse, there are various methods and devices that have been suggested as useful for encouraging loading. In the main these have been advertised in the non-scientific press, and have not been rigorously tested. The methods range from use of a demonstrator horse, through to specialised training techniques (discussed in Chapter 7). Devices such as electric prodders, hobbles, winches and specialist halters, rely on force and are inhumane. Other devices, such as the 'Easy-loader' (CAM Equestrian 1998), the 'Stableizer Restraint system' (a form of 'twitch' device, designed to work on acupuncture points) (Transcon Trading Co. Inc., 1997) and others such as covering the horse's head, may be useful, but are also likely to cause problems and have not been properly tested. The use of sedatives to aid loading has become common place, and can be useful in dealing with emergencies if they are used according to the instructions. If sedatives are to be effective, they must be given before loading has been started, whilst the horse is in a relaxed state. It must be remembered that sedation leads to reduced neuro-muscular control and so it is important that travelling is delayed until the horse has recovered. In most cases this will allow a horse to be loaded, although no learning about loading will have taken place. This means that the horse can become accustomed to the movement of the vehicle, restraint and confinement in those cases where it is unwilling to load without force.

4.2. CONFINEMENT AND ISOLATION

Once loaded into the vehicle, the horse is placed in a restricted space either due to being confined in an individual stall using partitions, or due to pressure on space by the rest of the group of loose horses with which it may be travelling. Horses are instinctively afraid of confinement due to their having evolved for life on the open-plain where being able to run away/escape was important for their survival (see Chapter 1). There has been little research into the possible benefits of windows in trailers and lorries. Horses possess good peripheral vision and therefore it is unclear where windows should be sited to ensure that they are used in a beneficial way. There are various differing opinions on whether windows are of benefit; some believe that they can create the illusion of space (Rees 1993), whilst some owners of horses, state that the shadows of moving cars and trees may cause a fear response. It is yet to be seen whether the provision of windows makes confinement more tolerable, but in the opinion of these authors (Waran & Friend, pers. obs.), most horses, when given a view, do appear to use it.

Confinement in the vehicle limits the behavioural opportunities available to the horse; in most cases they cannot turn around, move freely, or lie down. In addition they have very little control over their environment. The effects of confinement on the horse were investigated by Mal *et al.* (1991) when they stabled horses with and without companions and compared their responses to those of horses at grass. Their findings confirm that both isolation and confinement are stressful for horses.

The stall size is important and in some countries there is legislation in place that provides minimum space allowances for transporting different ages/types of horses by road (see AATA 2000, for information). Minimum space allowances differ from country to country (see Table 1).

Confinement within a specific space is usually achieved by means of partitions made of wood or metal, attached to the sides of the vehicle so that they can be moved/opened to enable the animal to be loaded and then shut in on both sides. Depending on the type of vehicle (trailer or lorry), and the internal layout, there may also be straps, bars or gates that enclose the horse's front and rear. The design of this internal furniture is crucial for safe and comfortable carriage of the horse, and yet there has been very little objective research in this area. For an average sized horse (height 158 cm, weight 550 kg) it has been suggested (Houpt and Lieb 1993) that a space of 90-cm wide, 2.4-m high and long should be used. In order to enclose the horse, this would require partitions to be of a length of 2.4 m, but the height and design of the partition is unclear. Many commercial vehicles are equipped with solid partitions that are as tall as the horse from the floor to the wither (full-height). There are other types of partitions that are solid from the wither or shoulder height to the top or middle of the leg, and then open down to the floor, sometimes with a rubber 'curtain' attached (half-height). An advantage of the half-

Table 1. Examples of space allowances (Floor area m^2/head) as required in Australia, Hungary and the EU.

Age of Horse (in months)	Country		
	Australia	Hungary	EU Directive
Adult	1.2	1.6–1.8[a]	1.75 (0.7 × 2.5)
18–24	1.0	–	1.2 (0.6 × 2) or 2.4 (1.2 × 2) if over 48 hours[c]
12–18	0.9	1.2–1.8[b]	1.4 (1 × 1.4)
5–12	0.7	–	

[a] In Hungary distinction is made between breeds/use of horse, 'warm-bloods' are given more space, and horses for slaughter less.
[b] Horses aged between foal and Adults are given increasingly more space.
[c] Under EU regulations horses of 12 to 24 months are considered together. Horses get more space if journeys are longer.

height partition is that it allows the animal to 'straddle' or 'brace' (position the front and hind legs wide apart), whilst the vehicle is moving. The height of the restraining bar, strap or gate to the front of the horse is especially important, since this breast-bar is designed to withstand the weight of the horse falling forward, and must also cause the least injury. Breast-bars or gates must therefore be adjustable, so that the stall can be flexible. In a study of 40 different sized horses transported in a commercial vehicle (Doherty & Waran 1994, unpublished data), it was found that the size of the horse significantly affected the position it took up in the stall and the frequency with which the ventral surface of the neck came into contact with the front bar. In this study all the horses were transported facing forwards and so the potential for damage during a sudden deceleration was high. Equally the same applies to the rump gate, especially if the vehicle is designed to carry the horses facing the rear of the vehicle. When facing the rear, it is essential that the rump-door or bar be positioned so that the horse can 'sit' on it, if the vehicle decelerates suddenly.

When horses are confined within a stationary vehicle for a short time, often the norm for many horses when attending a one-day competition in many countries, they appear to cope well as long as they are familiar with the process. Heart rates remain relatively low and similar to resting levels (Smith 1992; Waran & Cuddeford 1995) and horses appear to eat normally, stand in a 'normal' posture and sometimes rest a hind-leg (Waran & Cuddeford 1995). This relaxed attitude may be due to the horses in both of these studies being accustomed to the process, the climate was mild and they were with a companion. This would not necessarily be the case for young, feral or isolated horses, in parts of the world where temperatures can be extreme.

As a social species, the horse fears isolation. It takes some time before a horse will become accustomed to being separated from its stable/group mate and some horses are never able to accept this. Mal *et al.* (1991) observed that horses isolated in stalls vocalised frequently. Similarly, horses that are transported alone probably vocalise more than when they are transported with a companion. It is for this reason that many valuable performance horses will be stabled and transported with a companion that may or may not be of the same species (Coster 1996; Martin 1997). There are many anecdotal reports of the calming effect the familiar handler can have on a frightened horse. One author (Friend pers. obs.) has found transporting a single horse along with a person in the trailer to be a useful method of increasing bonding between the horse and people.

4.3. RESTRAINT

In addition to the restricted nature of the space available in a trailer or lorry, most sports/recreation horses will also be restrained within the individual stalls. Head ties are applied to prevent the horse from attempting to turn around, and for ease of handling. However, Stull (2001) found that even experienced horses are less stressed when their heads are not tied during transport. If the horse is tied too tightly this will prevent it from adopting the safest and most comfortable stance when the vehicle is in motion. This may lead to health problems due to the head being held too high, thus negatively affecting effective draining of the upper respiratory tract (Racklyeft & Love 1990). It is therefore important that the method of restraint is not too restrictive and allows the horse to move forward and backwards within the space available to it. The method of tying should also be considered, since if the horse does slip and go down, it must be able be able to get its head free of the restraint. It is for this reason that 'quick release' knots or similar mechanisms should be used for tying. When horses are transported in groups, where there are no individual partitions and where foals are transported with their dams, there should be no restraining devices used to avoid the horses becoming entangled.

4.4. THE MOVING VEHICLE

Once the engine has been started, the horse is exposed to a whole series of potential stressors. The engine noise, vibration of the vehicle, and the smell of the exhaust fumes will combine to provide a new challenge for the inexperienced traveller. Recorded sounds from a transport vehicle have been shown to induce the greatest increase in heart rates in pigs compared with their responses to the sounds on farm and at the abattoir (Talling *et al.*, 1996). However it was also shown that pigs habituate quite quickly to a consistent noise, and so this may only be stressful to animals during the first stages of transport. As the vehicle begins to move, the horse has to cope with additional problems such as staying upright (balancing), and coping with the environmental challenges related to being confined inside a semi-enclosed space (see Smith *et al.*, 1994, 1996)

4.5. ENVIRONMENTAL CHALLENGES

The space inside the moving vehicle is not usually conducive to a healthy environment. Good ventilation is vital for ensuring acceptable air temperatures, relative humidity and levels of contaminants such as gases and dust (Randall & Patel 1994). Despite the fact that the performance horse relies totally on a healthy respiratory system, we house and transport them in environments that may cause them respiratory problems.

Traub-Dargatz *et al.* (1988) demonstrated that following exposure to NH_3, NO_2 and CO, a horse's respiratory clearance is reduced. This is due to damage to the pulmonary epithelial barrier, which then becomes more permeable to bacteria, leading to an increase in the incidence of infection. Although no upper limit for exposure to ammonia (NH_3) and other gases has been recommended for equids, this should probably be lower than the limit laid down for humans, since horses have such a good sense of smell. In fact Cregier (1987) proposes that this level should be 20 ppm, based on a recommended limit of 25 ppm for humans laid down by the American Government Hygienists. As it is, Smith *et al.* (1996) found no adverse reactions in horses exposed to during transport levels of NH_3 less than 30 ppm. Thus exposure to such gases may not be a major problem in moving trailers (Stull 1999).

Ventilation of horse lorries and trailers has not been well researched (Cregier 1991), although studies of air flow around and within lorries used for transporting livestock (including hens), have been carried out (*e.g.* Randall 1993, Patel & Collins 1996). Tests using model horse lorries, carried out by Leadon and colleagues (unpublished data), provide valuable information on airflow around a horse. 'Puffs' or 'plumes' of air, usually enter the horse lorry via the windows or vents along the side of the lorry, and then this air drops towards the floor. At this stage the air becomes contaminated with dust from the bedding, gases and any bacteria. The air the horse inspires could potentially be of poor quality during these periods. However, during longer periods of transport, the dust is usually blown out and any noxious gases are usually well within acceptable ranges (Stull 1999). In trailers there is the additional possibility of the towing vehicle contributing to the poor air quality inside the trailer. Air vents are usually built into the trailer sides through which exhaust fumes may be drawn. Although legislation in most countries calls for vehicles used for transporting horses to be 'adequately' ventilated, the methods for achieving this are not fully understood.

Control of temperature and humidity may also be difficult in the horse transporter. Thermal stress occurs due to exposure to temperatures outside the relatively wide tolerance limits of the horse. Healthy horses have a wide temperature tolerance, being able to tolerate below freezing conditions up to desert conditions. Body temperature is a product of the balance between the metabolic heat produced and that lost to the environment via radiation, conduction, convection and evaporation. In the horse the main means by which heat is lost is via evaporation from the skin and the respiratory tract. Evaporation of sweat depends also on environmental humidity too, and it has been shown that high external temperatures coupled with a high humidity impair the evaporative process. Thus, horses in enclosed containers may experience heat stress if the internal temperature and humidity become excessive. However, because of the tremendous heat dissipation capacity of horses, conditions would have to be very hot and last for more than a short period of time. One author (Friend) has monitored the body temperature of horses when environmental temperatures reached 46 °C for an hour. The horses' body temperatures did not increase although the horses produced sweat profusely. Excessively high temperatures are likely where the ventilation system is poorly designed and ineffective. The potential for heat stress is greatest when the vehicle is stationary (*i.e.* no passive ventilation) and horses are under crowded conditions.

4.6. FOOD AND WATER INTAKE

Depending on the reasons for transportation, horses may or may not have the opportunity to feed and drink en route. Performance horses will often be offered forage in the form of hay or 'haylage' in a net, if the owner/manager feels that this will not impair performance. Horses habituated to transport will 'pick' at their forage, but it is likely that their intake will be reduced. The circumstances during transport appear to inhibit normal feeding and drinking behaviour probably resulting in hunger and thirst during and following transport and metabolic stress and electrolyte imbalance afterwards (Mars *et al.*, 1992; van den Berg *et al.*, 1998). Waran and Cuddeford (1995), found that horses in a moving vehicle fed less than when they were standing in a stationary vehicle, which suggests that the movement of the vehicle inhibits feeding. Hopes (1994) stated that many trainers of racehorses acknowledge that travelling has an effect on weight and fitness. When racehorses were transported by road for 6 hours and where food was offered, they still experienced an average 2.5% reduction in their body weight (Waran 1993). This weight was not made up until the third day after transport. Weight loss is likely to be due to a combination of defecation and urination while in the transport vehicle with reduced feed intake, reduced water intake and increased energy requirements. Foss and Lindner, (1996) also found that weight losses were related to distance travelled in a trailer, and dehydration is suggested to be an explanation for the weight losses found in the study by Smith *et al.* (1996). Reduced water intake is probably due to the horse's well-documented reluctance to drink water from an unaccustomed source (Mars *et al.*, 1992). In addition horses are often less willing to feed and drink in unfamiliar or stressful surroundings, or when distractions are present. Their motivation to escape and to avoid danger, perhaps over-riding their desire for food/water. However once horses become thirsty or dehydrated enough, they readily drink (Friend *et al.*, 1998; Gibbs & Friend 1999, 2000; Friend 2000). The effects of fear of novelty, can cause problems where the horse has been transported from its 'home' to a novel environment such as occurs for competition and race horses as well as horses transported to horse markets and abattoirs. Although dehydration was not well researched in transported horses, recent studies *e.g.* Loving (1997) showed that even a 12 hour journey can lead to loss of water through sweating and diahorrea, leading to at least 3% dehydration. Horses have the ability to go without water for extended periods of time during cool conditions. Tasker (1967) deprived horses of feed and water for up to 8 days during cool conditions at Cornell University in New York, US. These horses lost only 10% of their body weight by the end of the study. Carlson *et al.* (1979) deprived horses of feed and water under much warmer conditions, (11.8–33.2 °C) for 72 hours and observed a 10.7% loss in body weight. Dehydration rate is clearly accelerated under hot conditions when horses are required to sweat to maintain a consistent body temperature. Although the horses in the above studies were penned or stalled, recent research has shown that body weight loss of horses maintained in pens is similar to horses being transported (Friend et al., 1998; Friend 2000). Horses transported long distances during hot weather lost 8% after transport for 24 hours (Friend et al., 1998), 8% after transport for 30 hours when the night were relatively cool (Stull 1999) and 10.3% when transported for 30 hours under hot and humid conditions (Friend 2000).

If horses are provided with periodic watering during long distance transport after some initial weight loss, weight loss is stabilised during the trip. Dehydration, as measured by sodium and chloride concentrations, is delayed by periodic watering. Dehydration in non-watered horses, however, exceeds normal values by 8 hours of transport and continues to increase during the rest of a trip (Friend 2000). This high dehydration rate is the direct result of relatively high temperatures ranging between 24 °C during the night to 37 °C during the day. It is interesting to note that in this study (Friend 2000), when horses were penned in full sun and not watered, they dehydrated more rapidly than horses that were transported and not watered. This implies that the horses that were transported, experienced a slightly lower heat load.

Friend (2000) found that recovery from dehydration was relatively fast, with sodium and chloride returning to normal concentrations within a few hours of the horses being offered water. To reduce the incidence of mild colic from too rapid rehydration, horses that were as severely dehydrated as those in the study by Friend (2000), should not receive more than 6 litres of water at 30 min intervals. Although the horses that were offered water did not show severe dehydration, they showed severe signs of fatigue after 32 hours of transport necessitating the termination of transport. Although dehydration is not a major problem under cooler conditions, fatigue becomes the limiting factor in healthy horses transported for 28 hours or longer.

In performance horses on trips of moderate length, methods for increasing water intake include; offering familiar water and training horses to drink a flavoured water (Mars et al., 1992). It is important that the horse is offered and drinks water on arrival at its destination, and electrolyte supplements should be added to the water. When transporting performance horses, the ideal is to offer water whilst the vehicle is stationary (during breaks in the journey), at least every 2 to 4 hours especially when the external temperatures are high (Houpt & Lieb 1993). AATA guidelines (see Harris 2000), state that during transport, horses should be fed and watered every 15 hours and preferably every 6, depending on weather conditions. But it is also clear that consideration needs to be taken of when the animal was fed and watered prior to entering the transport vehicle. This is often problematic for horses bought through a market where the time spent in the market is unknown, and where feral horses sold in the market may be unwilling to drink the water available due to unfamiliarity and the method of acquiring it.

4.7. BODY POSTURE AND EVENTS DURING TRANSPORT

In order to remain upright when the vehicle is in motion, horses must make certain postural adjustments. Cregier (1981) observed that horses adopt a certain body posture whist being transported. The unstable surface that the horse must stand on appears to induce the horse to adopt a 'bracing' posture in a moving vehicle (Waran & Cuddeford 1995), where the hind-legs and fore-legs are held wide apart (splayed), the fore-limbs are advanced from the usual position beneath the body (Waran 1993) and at the head and neck are raised, so placing more weight over the hind-quarters (Cregier 1982; Roberts 1990). This posture has been observed amongst horses transported facing the direction of travel (forward facing), and at approximately 90 degrees to the direction of travel (slant or herring-bone) (Waran 1993). During transport, the horse is subjected to changing forces due, for example, to acceleration, movement of the vehicle around corners and deceleration. In addition the horse carries 60% of its body weight over its forelegs. The active response of the horse to transport is quite different from its normal body posture, where the legs are positioned directly beneath the body. The altered position of the legs during transit probably helps the horse to maintain its balance by allowing it to exert inclined thrusts with one leg or the other as occasion demands (Roberts 1990). However, although this may help the horse to remain upright during transit, it may also be physically tiring. Even during short journeys horses rarely shift from this position, and are therefore unable to relax whilst the vehicle is in motion. The slightly elevated heart rate associated with transport recorded by a number of researchers, may be associated with the effort needed to maintain balance in addition to general arousal. This is supported by Mars *et al.* (1992) who suggested that the elevated heart rates that were recorded during transportation of their horses in stock trailers were due to shifting of weight and stance to cope with highway driving. In addition, it has been suggested that the increase in heart rate measured in transported sheep is due to their need to brace themselves in order to remain upright (Baldock & Sibly 1990). Horses often lean (or are forced) against the partitions during a journey, which suggests that even the changes in their body posture during transit are not always adequate to counter all directional forces since this does not always stop them from over-balancing. Injury could occur to the horse and injuries associated with leaning have been reported in calves (Lambooy & Hulsegge 1988). Most animals experience bruising during transport (McCausland *et al.*, 1977). So although this is difficult to assess in live horses, it seems likely that is common. One method for protecting the horse against such injuries is by padding the internal partitions well. One company have taken this a stage further, designing air-filled ribs (Travel-safe Bumpsters™), that are attached to the walls/partitions to enable the horse to lean and to cushion the impact of any losses of balance. In preliminary tests (Plumbe 1998). These were shown to be used by horses travelling sideways (herring-bone), or travelling facing forwards or backwards in a lorry, when they lost balance. In more confined situations, perhaps where the partitions extend all the way to the floor (full height partitions), it is possible that horses would be unable to brace, would lose balance more easily and may be forced to use the partitions (and therefore any cushioning) more frequently. Montgomery (1992) demonstrated that horses were able to spread their fore and hind legs more easily where the partitions within the vehicle were half-height. Loss of balance and subsequent panic has been described where horses were given insufficient space to adopt the bracing stance (Tasker 1990). The road type also plays a role in determining how frequently the horse loses balance. Horses transported on minor roads, where there are more corners/bends in the road and an uneven surface, tend to lean against partitions more and show more leg movements (shifting position) than when on higher quality/faster roads.

4.8. ORIENTATION WITHIN THE VEHICLE

Various researchers have suggested that the main problem with transportation by road is related to the way in which the horse responds to the movements and vibrations of the vehicle (Tellington 1979; Cregier 1981; Roberts 1990). Particular concern has been expressed about the effects of sudden decelerations such as during braking or cornering, when the horse will be propelled towards the front of the vehicle. Cregier (1982) hypothesised that the horse is likely to be instinctively protective of its more vulnerable head and chest area and will hold its head and neck in an unnaturally high position as a means of protecting itself from injury. She also suggests that the hindquarters must continually shift and change position in order to counteract the changing forces imposed upon the horse when travelling in this position. This may explain why increases in plasma creatine kinase (CK, a muscle enzyme that leaks into the circulation if muscles have been traumatised) have been found in horses transported over distances of 130–200 km (reviewed in Leadon 1994). Interestingly, this was also found to be the case for transported deer, and in addition there was an effect of road type, with greater increases in CK associated with winding or minor roads (Grigor et al., 1998). However, consistent results have not always been obtained with CK. For example, CK was highest in the control horses that were penned for 30 hours and lowest in loose horses that were transported for 30 hours (Friend 2000). CK was also not useful in differentiating between control and loose horses that were transported for 24 hours (Friend et al., 1998). One suggestion for the raised heart rates generally associated with transportation, is that the horse is 'working hard' or expending energy whilst maintaining balance. In fact Doherty et al. (1997) demonstrated that ponies during transport experienced raised heart-rates and an increase in oxygen consumption (a measure of energy expenditure), as compared with when standing outside the lorry, standing in a stable and in a standing in a stationary vehicle. Interestingly they found that being transported caused the ponies to expend similar amounts of energy as during walking. It was suggested that energy expenditure and heart-rate changes could be used along with behavioural responses to determine the orientation within a moving vehicle most conducive to balancing effectively (Doherty et al., 1997). There was an indication from this study that the rear-face (backwards) position was associated with lower energy expenditure and heart-rates than if the ponies were facing the front of the vehicle or sideways (approx. 90 degrees from the direction of travel). This confirms the findings of various others (*e.g.* Slade 1987, Clark *et al.*, 1988; Smith 1994; Waran *et al.*, 1996). These authors claim that horses transported facing the rear of the vehicle, against the direction of travel, show fewer losses of balance, make contact with the sides of the lorry less, carry their head and necks lower and show a less rigid posture when they are facing the direction of travel (*i.e.* forward-facing). However the design of the vehicles used in these studies may also have influenced the results. Roberts (1990) claims that once the horse has learned how to sway 'passively' with the lateral movement of the vehicle, it can balance by rump-leaning and splaying its fore-legs. This is of particular advantage to horses when travelling in a trailer where the lateral forces due to the effect of the tow-hitch will be most significant. The possible reduction in physical effort by horses transported in the rear-face direction has been used to explain why some of the horses choose to position themselves facing away from the direction of travel when the vehicle is moving, although a number of horses also show a preference to face in the direction of travel (Smith *et al.*, 1994).

Although Crieger's hypothesis makes intuitive sense, research on orientation has found inconsistent results, perhaps related to differences in the design of trailers and horse boxes (floats, trucks) or preferences of certain horses. For example, in two-horse trailers, rear facing horses maintain their balance more easily (Clark *et al.*, 1993) and have lower heart rates and less movement than forward facing horses. However, the forward facing horses in this study were tied over a saddle compartment that greatly limited their head movement. Smith *et al.* (1994a) were unable to detect a difference in heart rates between horses facing forward or backward when transported in the back half of a four-horse 'stock' trailer. In another experiment in the same laboratory, horses spent significantly more time facing backwards when the trailer was in motion (Smith *et al.*, 1994b). However, some horses showed a very strong preference to face toward the direction of travel, and again, no differences in heart rates for either forward or backward orientations were evident. Interestingly, Kusunose and Torikai (1996) did find that yearling Thoroughbred horses increasingly oriented away from the direction of travel as the number of trials increased.

More recently, a series of experiments using a large trailer where simultaneous comparisons of different treatments (orientations) of horses (Gibbs & Friend 1999) transported four times over a course containing bumps, high speed turns and sudden stops was conducted. The results showed that orientation had only a slight effect. When individual horses were tied to the right side of the trailer, they spent 59% of the time facing the direction of travel. When tied on the left side of the trailer, the horses spent 52% of their time facing away from the direction of travel. The tendency to face forward when tied on the right and to face to the rear when tied on the left side of the trailer indicates a slight preference for the left side to be more exposed and the right side closer to the wall. The horses' ability to maintain balance when confined to stalls set at different orientations within the trailer was also determined in a series of trials. Horses facing backwards slipped more than horses that were forward facing, facing forward at a 45° angle, and rear facing at 45°, but leg movement and impacting or leaning on barriers or walls were not influenced by orientation. There was no significant difference between unshod and shod horses. Overall, a slight preference for a 45° orientation was observed, but no preference for facing either toward or away from the direction of travel was detected. In the opinion of these authors (Gibbs & Friend 1999), balancing ability was not meaningfully affected by orientation.

In another study by the same laboratory (Toscano & Friend 2001), orientation was again found not to have a significant effect upon the horse's ability to maintain balance over a set route. Perhaps the most interesting finding of that study was that certain horses showed much less movement in one orientation while others had much less movement in the opposite orientation. The authors conclude that if facing forward is adverse, horses would be expected to spend little time in that orientation, but the research data shows only a slight preference, at best, for facing away from the direction of travel (see Figure 1).

The strong preference that some horse owners have observed for horses to voluntarily face away from the direction of travel may be due to various factors. These include; an improved ability to maintain balance in some individuals, avoidance of saddle compartments/racks that may restrict the movement of the head and neck, preference to avoid the dark-cave effect associated with the forward end of most smaller horse trailers and a strong desire to face the more open rear of a trailer that often has a large opening between the top of the rear doors and the roof of the trailer. Observations from a commercial horse transport company that operates large vans in which horses face both toward and away from the direction of travel indicate that having horses tied in such a manner that they can readily raise and lower their heads is the most important factor in reducing transport stress (Robert Maxwell, personal communication). Stull and Rodiek (submitted) found that horses that were not cross tied during transport showed fewer signs of stress (cortisol, glucose, white blood cell count and neturophil to lymphocyte ratio) than horses that were transported cross tied in the same van.

4.9. DENSITY

Most horse owners are not concerned about the number of horses placed in a trailer because competition, performance and companion horses are transported in individual stalls. However, density is a very important issue for horses being transported in loose groups. Many truck drivers will tell you that cattle and horses 'hold each other up' when they are packed tight on the trailer. People in the cab of a tractor that is pulling a large trailer loaded with loose cattle or horses can hear and feel the animals in the trailer as the animals impact the floor and sides of the trailer and there is less banging around when the animals are tightly loaded. However, recent studies with horses and cattle have given new insight to that claim.

In lower density situations, horses have more opportunity to move a hoof or change their posture to compensate for changes in speed and hence there are more body and hoof impacts that the driver hears (Collins *et al.*, 2000). This concurs with Tarrant and Grandin's (1993) conclusion that shifting within trailers is inhibited at high stocking density and that there is a corresponding increase in struggles and falls at high stocking density. In addition, the ability of horses to stand up after a fall was clearly hampered by high density (Collins *et al.*, 2000), which contributed to the greater number and greater severity of injuries observed in those horses (Collins *et al.*, 2000). When horses fell at a high stocking density, they were trapped on the floor by the remaining animals closing over and occupying the available standing space (Collins *et al.*, 2000). In the controlled studies conducted by Collins *et al.* (2000), horses that fell in a high density treatment spent a much longer time down than those transported in the lower stocking density. Although those horses were transported for just 25 min, one horse in the third trial ceased attempting to

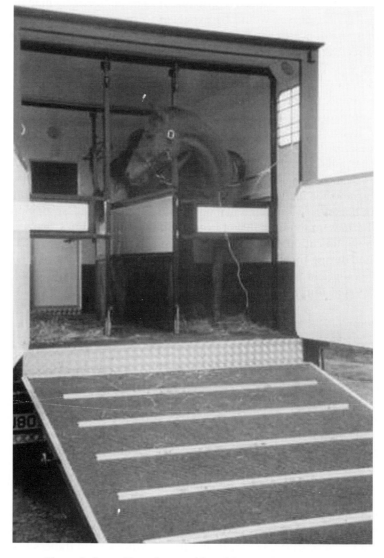

Figure 1. Lorry Horse-box partitioned for rear-face transport.

get up shortly after falling and had to be rescued. Most horses that go down under high-density conditions during long distance transport will be trampled if aid is not rendered quickly.

The increased number and severity of injuries in high-density conditions may be due to a number of factors. Experienced horses avoid contacting surfaces during transport (Gibbs & Friend 2000) and prefer to maintain their balance independently of other horses. Horses may be crowded to the point that they can move only several inches and are forced to have contact with other horses and the sides of trucks (Collins et al., 2000). They constantly reposition their feet in an apparent attempt to maintain their balance and frequently step on the hooves and pasterns of other horses. Aggressive horses bite and kick other horses, resulting in injuries as they manoeuvre for room. It only takes one or two aggressive horses in a group to disrupt a trailer load of horses (Grandin et al., 1999; Collins et al., 2000). The contribution to injuries made by an aggressive horse is also amplified by the attempts of horses being bitten to stop the biting by kicking with their hind feet. Because of the crowding, the bitten horses cannot move away and invariably kick the wrong horse. Injuries, of course, may also be influenced by the type of flooring, wet manure and other factors affecting a horse's traction.

High stocking densities create a situation of constant struggle for the horses. Decreasing density would reduce the overall stressfulness of long distance transport by allowing the horses some manoeuvring room to avoid aggressive horses, to stand in a more comfortable position, to adopt their preferred orientation, and perhaps allow them to rest during periods when the truck is stopped (Collins et al., 2000). Reducing stocking density may reduce injury and bruising during transportation, but will increase transport costs.

Choice of the best orientation for transporting horses will be dependent upon the horse's experience, its preference and the method of transportation. In addition many owners are constrained by the existing design of their vehicle. It is not as easy as reversing horses into conventionally built forward-facing trailers. The interior, the axles, and the tow bar/hitch must all be properly adjusted to ensure that the trailer can cope with the different weight distribution (Foster 1981).

5. Effects of Transport on Performance

Much of the work in this area is conflicting or inconclusive, for example, early embryonic death in pregnant mares being transported was reported by van Neirkerk and Morgenthal (1982), but Baucus et al. (1989) did not find this to be the case. Despite many horses being transported specifically for performance purposes, surprisingly little is known about the impact transportation has on performance. There are a number of reports that suggest that for experienced horses, transport over short distances has little impact on performance (e.g. Beaunoyer & Chapman 1987). When Slade (1987) investigated the effects of rear and forward facing transport on post transit racing performance, no effect was found. The same seems to be the case for horses used for show-jumping, although less experienced horses did exhibit signs of stress (Covalesky *et al.*, 1991; Russoniello *et al.*, 1991). However when the effects of longer journeys have been looked at, there are indications that the response to transport could compromise racing performance (Linden *et al.*, 1991). There is also the matter of the sort of exercise that is required of the horse following transport. Linden *et al.* (1991) found that certain types of exercise induced different cortisol responses, and this must be taken into account when considering the impact that transportation may have. There is also the question of individuality. Some horses cope better with transport than others, and recovery will be dependent upon the health and fitness of the individual, the transport experience and other factors such as road type and driver technique. One of the problems in studying this area is being able to develop good scientific methodology for assessing the effect of transport on performance. Physiological and biochemical measures of stress (*e.g.*, raised plasma cortisol levels, increased CK, increased heart-rate) are often inconclusive since they need to be collected (which may in itself cause a stress response), may be clinically insignificant (Leadon 1994), and can usually only be measured before and after transport, which may mean that the changes have been missed. What does seem clear is that very little is known about the relationship between transport and performance and about ways of reducing the impact of transport.

6. Air Transport

The size and extent of the international horse air transport industry is seldom grasped by those unfamiliar with it. However, an illustration of the size and rate of growth of this industry can be gained from measuring dual hemisphere transport which was initiated by Coolmore Stud with Godswalk in the late 1980s. Today, as many as 100 stallions from the world's leading studs may be mated with mares in both the Northern and Southern Hemispheres in the same year. They travel, as do so many other horses, between Ireland, the UK and France and also from the USA and Japan, Australia and New Zealand. Horses also travel by air to and from North and South America for breeding and racing and stallions' owners are now planning to send their stallions from Europe to Argentina and Brazil.

Quarantine regulations are applied to this travel and trade. These may vary from country to country, but are always intended to prevent the spread of disease. Many horse populations are, in disease terms, naive. Equine influenza was common and caused major problems in South Africa in the 1980s. Ten years ago, this disease was estimated to have the potential to cause losses of $220 million per month, if it reached Australia. The recent Foot and Mouth Disease crisis in the UK (2001) resulted in additional pre-export safeguards and necessitated a quarantine period for UK horses in Ireland, prior to their export to Australia. Stallions and mares completed their quarantine in strict isolation at approved facilities on private and public stud farms in Ireland, France and in the USA. One quarantined, they are cleared for export, under government supervision and travel by road to various airports in a variety of vehicles. The entire and complex process is organised and supervised by bloodstock transport specialists who arrange and supervise the transfer of horses from door to door.

On arrival at the airport, horses are transferred to air stables or jet stalls (see Figure 2). They are transported in a wide bodied jet, like the Singapore Airlines Boeing 747 jumbo jet used to carry mares from Dublin to Sydney or the Fedex operated MD11 jet, which transported USA based stallions from Louisville, Kentucky to Australia. The evolution of the jet stall has revolutionised the international transport of horses in the last 10 years or so. 'Jet stall' arrangements have today, largely superseded the previous 'open stalls', other than in the narrower bodied, older aeroplanes, many of which are still in use or on the short haul aircraft used to transport racehorses around Europe and elsewhere. Designed to confine horses in safety at altitude, jet stalls facilitate the rapid loading and unloading which is an essential feature of all large scale modern air cargo activity. These stalls have been designed to accommodate a maximum of three horses, side by side, separated by partitions. Access to the horses for feeding and watering and for their expert travelling grooms is provided at the front of the stalls. Flying grooms are highly experienced professionals, who in addition to their horse skills have certified expertise in air safety (see Harris 2000). Horses can for a higher price, also travel two to a stall or even enjoy a stall entirely to themselves, if economics allow.

The jet stalls once loaded and cleared by the authorities are then transferred to the side of the aeroplane. Tugs or tractors, take one or more jet stalls in little trains to where they are rolled onto a scissors lift that raises them and passes them into either the front, or the rear side door of the aeroplane. The positioning of jet stalls within the aircraft, (other than ensuring colts and stallions are placed in front of

Figure 2. A Jetstall for transport of horses by air.

fillies or mares) is determined by Loadmasters. They are accredited experts who ensure that all of the freight on board is loaded correctly and that is safely secured, in accord with the weights and balance distributions specified by the aircraft manufacturer. Jet stalls may be arranged in either a single row or in two parallel rows on the main cargo deck. A maximum of 29 jet stalls containing 87 horses can be carried by a 747 Jumbo jet. These jets seldom carry all-horse loads. Other freight is often carried in addition to the horses. The huge 747 – 200 series freighter can carry a load of 200,000 lbs over a range of 4490 miles.

Horses can therefore travel to many international destinations without a stop for re-fuelling en route. However, travel to extreme distances requires one or more re-fuelling stops and loading and unloading of other freight may take place simultaneously with refuelling. Horses travelling to Australia and New Zealand from Ireland, the UK or France usually travel eastwards and may have re-fuelling/loading and unloading stops in Continental Europe, the Middle East and Singapore. Insect proof netting must be fitted to the jet stalls if horses have to change planes in Singapore, to prevent the introduction of insect-borne diseases.

Most horses tolerate confinement, jet noise and repeated take-offs and landings well. Their dedicated handlers offer them water every 4 to 6 hours throughout their journey and hay or haylage is usually made available to them throughout their flight. Veterinary care is provided where economics permit. Although some lose no weight during their journey it is not unusual for horses to lose about 20 kilos or 4% of their bodyweight on a 24 hour journey. Some horses, can suddenly and unpredictably resent confinement. This occurs in less than half of one per cent of horses that travel. Caring attention and re-assurance, physical restraint and access to modern indictable tranquillisers are generally sufficient to restore calm. Extreme frenzy can be almost impossible to control in such a confined space and has, very rarely, resulted in fatality.

The rarity of frenzy related deaths are a considerable tribute to the skill of the flying grooms. Their prompt intervention has saved many lives. Like their human counterparts, a minority of equine passengers may develop aviation-related disease. Although colic and other illness may manifest itself at altitude as it can anywhere else, the principal problem associated with the long distance transport of horses is respiratory disease. Known colloquially as 'Shipping Fever' this disease has been recognised for over 100 years, since horses had to travel to many of today's destinations by sea. The disease had, until relatively recently, an incidence of about 6%, *i.e.* 94% of horses transported long distances were unaffected by the combination of pneumonia and pleurisy that is 'Shipping Fever'. The clinical signs of this shipping fever include depression, reluctance to drink, increased rectal temperature, and increased respiratory rate with other signs of respiratory disease. Research by Raidal *et al.* (1985, 1996, 1997) has shown that holding a horses head high, for protracted periods, even on the ground can pre-dispose to the development of this disease. In addition, studies funded by the International League for the Protection of Horses and the International Equestrian Federation, and carried out by the Irish Equine Centre, have shown that environmental conditions on aircraft can contribute to the development of respiratory disease (*e.g.* Leadon 1989; Leadon *et al.*, 1990, 1994). For example, bacterial numbers in the air in the cargo hold can increase dramatically during the journey especially at re-fuelling stops.

Data loggers are now placed in jet stalls and temperature and relative humidity can be measured digitally throughout a long air journey. Although still in their early phases, these studies are illustrating the extreme variations in air moisture that can occur. Very dry air can desiccate the respiratory system and interfere with clearance from the lungs. These studies are potentially important because the incidence of shipping fever is now higher than when open stalls were normal. A total of 20% or more of horses transported long distances in jet stalls can show fever during the journey or thereafter. Modification of jet stall design may be required to reduce the incidence of this disease.

The provision of veterinary care at altitude has been shown to have significant equine welfare benefits. Trauma and frenzy can be dealt with as they arise, the very rare episodes of colic can be addressed and the duration and severity of shipping fever is much reduced as a result of diagnosis and treatment in the air. Although this section relates to air transport, shipping fever is also seen after road transport. Careful checking of horses after arrival following long journeys is essential. Taking of rectal temperatures twice daily for at least three days after arrival will identify shipping fever cases, since this is probably one of the single most important welfare measures in the transport context. Prompt veterinary attention in any instance of fever after shipping, with the advice that any affected horse has just completed a long journey, is essential.

7. Conclusion

The transport of horses is a complex issue that may have significant effects on horse welfare if not carried out correctly. Recently there has been a considerable amount of research, particularly on the transport of horses by road. The results of the studies described in this chapter need to be available to the horse owning public in order for the necessary changes to be made to practice and legislation.

8. References

AATA (2000) *Manual for the Transportation of Live Animals*, 2nd edition. T.C. Harris, Redhill, UK 156 pp.

American Horse Council (1996) *The Economic Impact of the Horse Industry in the United States, Vol. 1, National Summary*. American Horse Council Foundation report prepared by Policy Economics Practice, Barents Group LLC.

Baldock, N.M. and Sibley, R.M. (1990) Effects of handling and transportation on the heart rate and behaviour of sheep. *Applied Animal Behavior Science* **28**, 15–39.

Barker, T.C. (1983) The delayed decline of the horse in the twentieth century. In Thompson, F.M.L. (ed.), *Horses in European Economic History: A Preliminary Canter*. British Agricultural History Society, Reading, UK.

Baucus, K.L, Ralston, S.L., Nickels, C.F., McKinnon, A.O. and E.L. Squires (1990) Effects of transportation on early embryonic death in mares. *J. Animal Science* **68**, 335–345.

Beaunoyer, D.E. and Chapman J.D. (1987) Trailering stress on subsequent submaximal exercise performance. *Proceedings of 11th Equine Nutrition and Physiology Symposium.* Oklahoma State University, Stillwater, Oklahoma, USA, pp. 379–384.

Board of Trade (1915) *Annual Railway Returns.* HMSO, London, UK.

British Horse Society (1998) *The BHS Complete Manual of Stable Management.* The British Horse Society, Warwickshire, UK.

Bush, K. (1992) *The Problem Horse – An Owners Guide.* Crowood Press, UK.

Cam Equestrian (1998) *Trade Literature.* Cam, Eardisley, UK.

Carlson, G.P., Rumbaugh, G.E. and Harrold, D. (1979) Physiologic alterations in the horse caused by food and water deprivation during periods of high environmental temperatures. *American J. Veterinary Research* **40**, 982–985.

Clark, D.K., Friend, T.H. and Dellmeier, G. (1993) The effect of orientation during trailer transport on heart rate, cortisol and balance in horses. *Applied Animal Behaviour Science* **38**, 179–189.

Collins, M.N., Friend, T.H., Jousan, F.D. and Chen, S.C. (2000) Effects of density on displacement, falls, injuries, and orientation during horse transportation. *Applied Animal Behaviour Science* **67**, 169–179.

Coster, G. (1996) It's a long way to Epsom. *Independent on Sunday,* 2 June, p. 73.

Covalsky, M., Russoniello, C. and Malinowski, K. (1991) Effects of show-jumping performance stress on plasma cortisol and lactate concentrations and heart rate and behaviour in equines. *Proceedings of 12th Equine Nutrition and Physiology Symposium.* University of Calgary, Canada, pp. 1171–1172.

Cregier, S.E. (1982) Reducing equine hauling stress. A review. *J. Equine Veterinary Science* **2**, 186–198.

Creiger, S.E. (1987) The psychology and ethics of humane equine treatment. In Fox, M. and Mickley, L.D. (eds.), *Advances in Animal Welfare Science 1986/87.* Martinus Nyhoff, Netherlands.

Creiger, S.E. (1991) Notes for an interview with Horse Illustrated, Personal Communication.

Doherty, O., Booth, M., Waran, N.K., Salthouse, C. and Cuddleford, D. (1997) Study of heart rate and energy expenditure of ponies during transport. *Veterinary Records* **141**, 589–592.

Forhead, A.J., Smart, D. and Dodson, H. (1990) Transport induced stress responses in the donkey. *J. Endocrinology, Suppl.* **127**, 91.

Foss, M.A. and Lindner, A. (1996) Effects if trailer transportation duration on body weight and biochemical variables of horses. *Pferdeheilkunde* **12**, 435–437.

Foster, C. (1981) Travelling the horse. *Equine* Nov/Dec, 21–23.

Friend, T.H. (2000) Dehydration, stress, and water consumption of horses during long-distance commercial transport. *J. Animal Science* **78**, 2568–2580.

Friend, T.H. (2001) A review of recent research on the transportation of horses. *J. Animal Science Suppl.* **79**, E32–40.

Friend, T.H., Martin, T.M., Householder, D.D. and Bushong, D.M. (1998) Stress responses of horses during a long period of transport in a commercial truck. *J. American Veterinary Medical Association* **212**, 838–844.

Gibbs, A.E. and Friend, T.H. (2000) Effect of animal density and trough placement on drinking behaviour and dehydration in slaughter horses. *J. Equine Vet. Sci.* **20**, 643–650.

Gibbs, A.E. and Friend, T.H. (1999) Horse preference for orientation during transport and the effect of orientation on balancing ability. *Applied Animal Behaviour Science* **63**, 1–9.

Gibbs, P.G., Potter, G.D., Jones, L.L., Benefield, M.R., McNeill, J.W., Johnson, B.H. and Moyer, W. (1998) *The Texas Horse Industry.* Texas Agricultural Extension Service, Texas A&M University System, College Station, Texas, p. 15.

Gibbs, P.G., Benefield, M.R., Potter, G.D., McNeill, J., Johnson, B.H. and Moyer, W. (1997) *Profile of Horse Ownership and Use in 8 Texas Counties – Phase 2 of the Texas Horse Industry Quality Audit.* In Proceedings of the 15th Equine Nutrition and Physiology Society, Texas, USA, pp. 321–325.

Grigor, P.N., Goddard, P.J., Littlewood, C.A. and MacDonald, A.J. (1998) The behavioural and physiological reactions of farmed deer to transport: effects of road type and journey length. *Applied Animal Behavioural Science* **56**, 263–279.

Harris, T. (29 March 1996) Council of Europe (Email to P. Norris) (Online) Available email: 100257.1720@compuserve.com

HMSO (1822) *An Act to Prevent the Cruel and Improper Treatment of Cattle*, 22.7.1822.

Hopes, R. (1984) The balance of welfare. *Equine Veterinary J.* **16, 1**, 1–3.

Houpt, K.A. (1982) Equine behaviour. *Equine Practice* **4, 2**, 12–16.

Houpt, K.A. and Leib, S. (1993) Horse handling and transport, In Grandin, T. (ed.), *Livestock Handling and Transport*. CAB International, UK.

Kusunose, R. and Tonkai, K. (1996) Behaviour of untethered horses during vehicle transport. *J. Equine Science* **7**, 21–26.

Larter, C. and Jackson, T. (1987) *Transporting your Horse and Pony*. David and Charles, Newton Abbot, UK.

Leadon, D., Daykin. J., Blackhouse, W., Frank, C. and Attock, M.A. (1990) Environmental haemato-logical and blood biochemistry changes in equine transit stress. *Proceedings of American Association of Equine Practise* **36**, 485–490.

Leadon, D., Frank, C. and Blackhouse, W. (1989) A preliminary report on studies on equine transit stress. *J. Equine Veterinary Science* **9, 4**, 200–202.

Leadon, D.P. (1994) Transport stress. In Hodgson, D.R. and Rose, R.J. (eds.), *The Athletic Horse: Principles and Practise of Equine Sports Medicine*. W.B. Saunders Co., UK, pp. 372–378.

Linden, A., Art, T., Amory, D., Desmecht, D. and Lekeux, P. (1991) Effect if 5 different types of exercise, transportation, and ACTH administration on plasma cortisol concentration in sport horses. *Equine Exercise Physiology* **3**, 391–196.

Loving, N.S. (1997) *Go the Distance*. Kenilworth Press, Buckingham, UK.

Mal, M.E., Friend, T.M., Lay, D.C., Vodelsang, S.G. and Jenkins, O.C. (1991) Physiological responses of mares to short-term confinement and isolation. *Journal of Equine Veterinary Science* **11**, 96–102.

Marlin, D. (1995) Horses under heat stress. In *International Conference on Feeding Horses-scientific Session*. Dodson and Horrell Ltd., Kettering, UK.

Mars, L.A., Keisling, H.E., Ross, T.T., Armstong, J.B. and Murray, L. (1992) Water acceptance and intake in horses under shipping stress. *Equine Veterinary Science* **12**, 17–21.

Martin, B. (1997) That animal magnetism. *Daily Telegraph*, 12 April.

McCausland, I.P., Austin, D.F. and Dougherty, R. (1977) Stifle bruising in bobby calves. *New Zealand Veterinary J.* **25**, 71–72.

Montgomgery, S. (1992) Finding hidden stresses if a bad traveller. *Racing Post*, 8 October.

Owen, R.R., Fullerton, J. and Barnum, D.A. (1983) Effects of transportation, surgery and antibiotic therapy in ponies infected with Salmonella. *American J. Veterinary Research* **44**, 46–50.

Patel, K. and Collins, J. (1996) Specialist livestock vessels. *State Veterinary J.* **6**, 3 October, 1–4.

Phillips, P.A., Thompson, B.K. and Fraser, D. (1988) Preference tests of ramp design for young pigs. *Canadian J. Animal Science* **68**, 41–48.

Plumbe, A.E. (1998) *Physiological and Behavioural Responses of Horses to Road Transportation*. MSc Thesis, Institute of Ecology and Resource Management, University of Edinburgh.

Racklyelf, D.J. and Love, D.N. (1990) Influence of head posture on the respiratory tract of healthy horses. *Australian Veterinary J.* **67, 11**, 402–405.

Raidal, S.L. Love, D.N. and Bailey, D.G. (1996) Effects if posture and accumulated airway secretions on tracheal mucociliary transport in the horse. *Australian Veterinary J.* **73**, 433–438.

Raidal, S.L., Love, D.N. and Bailey, D.G. (1995) Inflammation and increased numbers of bacteria in the lower respiratory tract of horses within 6 to 12 h of confinement with thte head elevated. *Australian Veterinary J.* **72**, 45–50.

Randall, J.M. (1993) Environmental parameters necessary to define comfort for pigs, cattle and sheep in livestock transporters. *Animal Production* **57**, 299–307.

Randall, J.M. and Patel, R. (1994) Thermally induced ventilation of livestock transporters. *Agricultural Engineering Research* **57**, 99–107.

Rees, L. (1993) *The Horses Mind*. Stanley Paul Ltd., London, UK.

Roberts, T.D.M. (1990) Staying upright in a moving trailer. *The Equine Athlete* **3**, 2–8.

Russoniello, C., Racis, S.P., Ralston, S.L. and Maliniwski, K. (1991) Effects of show-jumping per-formance stress in haematological parameters and cell-mediated immunity in horses. *Proceedings of 12th Equine Nutrition and Physiology Symposium*. University of Calgary, Canada, pp. 145–147.

Slade, L.M.J. (1987) *Trailer Transportation and Racing Performance.* Proceedings 10th Equine Nutrition and Physiology Symposium, 11–13 June, pp. 511–514. Fort Collins, CO, Texas, USA.

Smith, B.L. (1992) Minimising transport stress in horses. *Equine* Feb, 14–15.

Smith, B.L., Jones, J.H., Carlson, G.P. and Pascoe, J.R. (1994a) Effect of body direction on heart rate in trailered horses. *American J. Veterinary Research* **55**, 1007–1011.

Smith, B.L., Jones, J.H., Carlson, G.P. and Pascoe, J.R. (1994b) Body positions and direction preferences in horses during road transport. *Equine Veterinary J.* **26**, 374–377.

Smith, B.L., Miles, J.A., Jones, J.H. and Willits, N.H. (1996b) Influence of suspension design on vibration in a two-horse trailer. *Trans American Society of Agricultural Engineering* **39**, 1083–1092.

Smith, B.L., Jones, J.H., Hornof, W.J., Miles, J.A., Longworth, K.E. and Willits, N.H. (1996a) Effects of road transport on indices of stress in horses. *Equine Veterinary J.* **28**, 446–454.

Stephens, D.B., Bailey, K.J., Sharman, D.F. and Ingram, D.C. (1985) Analysis of some behavioral effects of vibration and noise components of transport in pigs. *Quarterly J. Experimental Physiology* **70**, 211–217.

Stull, C.L. (1999) Responses of horses to trailer design, duration, and floor area during commercial transport to slaughter. *J. Animal Science* **77**, 2925–2933.

Stull, C.L. (2001) Evolution of the proposed federal slaughter horse transport regulations. *J. Animal Science* **79**, E12–E15.

Stull, C.L., and Rodiek, A.V. (submitted) Effects of cross-tying horses during 34 hours of road transport. *Equine Veterinary J.*

Talling, J.C., Waran, N.K., Wathes, C.M. and Lines, J.A. (1996) Behavioral and physiological responses of pigs to sound. *Applied Animal Behaviour Science* **48**, 187–201.

Tarrant, V. and Grandin, T. (1993) Cattle transport. In Grandin, T. (ed.), *Livestock Handling and Transport.* CAB International, Wallingford, UK pp. 109-126.

Tasker, J.B. (1967) Fluid and electrolyte studies in the horse. IV. The effects of fasting and thirsting. *Cornell Veterinary J.* **57**, 658–667.

Tasker, W.J. (1990) Transport problems. *Equine Behaviour* **25**, 19–20.

Tellington, W.J. (1979) *Endurance and Competitive Trail Riding.* Doubleday & Co., New York, USA.

Tradesman Industries Inc. (1995) *Video and Trade Literature.* California, USA.

Traub-Dargatz, J.L., McKinnon, A.O., Bruyninckx, W.J., Thrall, M.A., Jones, R.L. and Blancquaert, A-M.B. (1988) Effect of transport stress on bronchoalveolar lavage fluid analysis in female horses. *American J. Veterinary Research* **49**, 7, 1026–1029.

Van den Berg, J.S., Guthrie, A.J., Meintjes, R.A., Nurton, J.P., Adamson, D.A., Travers, C.W., Lund, R.J. and Mostert, H.J. (1998) Water and electrolyte intake and output in conditioned Thoroughbred horses transported by road. *Equine Veterinary J.* **30**, 4, 31–323.

Van Niekerk, C.H. and Morgenthal, J.C. (1982) Foetal loss and effects of stress in plasma progesterone levels in pregnant mares. *J. Reproduction and Fertility, Suppl.* **32**, 453–457.

Von Mayr, A. and Siebert, M. (1990) Studies of the paraspecific immunostimulant PIND-ORF on transport induced stress in horses. *Tierarztliche Umschau* **45**, 677–682.

Waran, N.K. (1993) The behaviour of horses during and after transport by road. *Equine Veterinary Education* **5**, 129–132.

Waran, N.K. and Cuddeford, D. (1995) Effects of loading and transport on the heart rate and behaviour of horses. *Applied Animal Behaviour Science* **43**, 71–81.

Waran, N.K., Robertson, V., Cuddleford, D. Kokoszko, A and Marlin, D.J. (1996) Effects if transporting horses facing either forwards or backwards on their behaviour and heart rate. *Veterinary Records* **139**, 7–11.

Warriss, P.D., Bevis, E.A., Edwards, J.E., Brown, S.N. and Knowles, T.G. (1991) Effects of an angle of slop on the ease with which pigs negotiate loading ramps. *Veterinary Record* **128**, 419–4421.

White, A., Reyes, A., Godoy, A. and Martinez, R. (1991) Effects of transport and racing on ionic changes in Thoroughbred race horses. *Comparative Biochemical Physiology* **99a**, 3, 343–346.

Chapter 7

TRAINING METHODS AND HORSE WELFARE

N. WARAN
Department of Veterinary Clinical Studies, Royal (Dick) School of Veterinary Studies, University of Edinburgh, Easter Bush, Roslin, Midlothian EH25 9RG, UK

P. McGREEVY
Faculty of Veterinary Science, Gunn Building (B19), Regimental Crescent, University of Sydney, NSW 2006, Australia

R.A. CASEY
Anthrozoology Institute, School of Biological Sciences, University of Southampton, Bassett Crescent East, Southampton, Hampshire, UK

Abstract. Many aspects of horse care and handling are based upon convenience and traditional practices. Many of these methods of management and practice do not take into account the natural behaviour of horses. This is despite the belief that although domestic horses are probably more docile, stronger, faster growing and faster moving than their ancestors, they are unlikely to have lost any natural behaviours. The performance or sport horse is expected to perform a wide variety of movements and tasks, some of which are unnatural or exaggerated and most of which must be learned. The term 'training' is commonly used to describe the processes whereby the human handler introduces the horse to new situations and associations. Performance horses are often required to tolerate stimuli that are innately aversive or threatening, such as having a person on their backs. They are also trained to respond to a stimulus with often unnatural or over-emphasised behaviour, such as some of the dressage movements. Effective and humane training requires an understanding of the processes underlying behaviour. These include knowledge of behaviour under natural conditions, learning processes, the influence of early experience and motivational forces. Horses differ from the other main companion animal species, namely cats and dogs, in that they are a prey species. They most commonly flee from dangerous and painful situations. Horses readily learn to avoid potentially threatening situations and if their attempts to avoid associated stimuli are prevented, they will often exhibit problem behaviours. In this chapter the history of horse training, the application of learning theory and a knowledge of equine behaviour to training, and innovative training methods are all considered.

1. Introduction

The quest for gentle approaches to training horses has been a conscious aim only over the course of the last century in Northern Europe and America. This upsurge in interest in training methods that utilise the natural behaviour of the horse may be linked as much to the increased wealth of these countries, that has led to more people being able to afford to keep horses purely for pleasure, as to the improved scientific awareness of animal feelings and consciousness. By contrast, in the less developed countries, attitudes to welfare are necessarily different, with equids being

N. Waran (ed.), The Welfare of Horses, 151–180.
© 2002 *Kluwer Academic Publishers. Printed in the Netherlands.*

considered in utilitarian rather than affectionate terms – people's concerns being primarily based on their own welfare and that of their families.

Much of the philosophy of horse training that commonly pervades the horse industry is based on traditional practices that have developed since the domestication of the horse. These traditional training techniques had some degree of effectiveness, but were not necessarily designed with the welfare of the horse in mind. In order to understand why certain training practices exist, we have to consider the history of taming and training horses, and the perceived benefits of prevalent techniques. It is appropriate to evaluate these training techniques in the light of current attitudes to the scientific appraisal of animal welfare.

2. The History of Training

Historically attitudes towards horses were fundamentally utilitarian. Throughout most of the history of their relationship with people, horses have represented food sources or facilitators of food procurement, and were present in sufficient numbers to be considered expendable. The earliest interactions between man and horse was as hunter and prey animal, as evidenced by cave paintings by Cro-Magnon humans over 15,000 years ago, and the piles of equid bones found near human settlements. As groups of people became more sedentary it was appropriate for them to devise methods of keeping horses near to their settlements rather than allowing them to range freely. The first handling or 'taming' of horses was to enable people to use them as a 'milk and meat' source. There is some archaeological evidence that horses started to be ridden around 6000 years ago (*e.g.* Telegrin 1986) in the Ukraine – some time before they were first used as draught animals in North Africa (Clutton-Brock 1992). The main evidence of human intervention is from the skulls of horses found at the settlements, in which tooth wear was comparable with that found in horses ridden in a metal bit. To produce herd replacements in the absence of domesticated stallions, tame mares may have been restrained by fibre or hide hobbles outside the settlements. Whilst many of the resultant foals were used for meat, the mares would have been milked – a practice developed previously with cattle (Clutton-Brock 1987).

Clues as to early methods of capturing and restraining horses were available from the Mongols, who retained a 'horse economy' right up until the last century. Horses were caught by wounding them in a hunt, by chasing mares and foals for several days until the latter were exhausted, or through the use of traps and snares, natural ravine 'corals', and tame decoys (Clutton-Brock 1992). To employ such methods, tribesmen would have had a good understanding of the herding behaviour of horses to predict their movements, and direct them to where they wished them to go.

With time, the purposes of horse-keeping and the sophistication of horse husbandry and management increased. For example, as greater control was achieved through the use of twisted horsehair headcollars and tethers, it seems likely that horses were used to drag travois (heavy loads) or carry the old or infirm from one camp to the next. As foals that were born to the tethered mares grew, the herdsmen were faced with having to handle younger and fitter animals, instead of the kinds of horses they had previously caught in corals. Hence, the headcollar probably developed further to enable the horsemen to have better control of the horses' heads. As the herds became larger horse tethering became more onerous and containment became logistically more difficult. It seems likely that these problems prompted herdsmen to mount their horse in order to control their stock.

There is pictorial evidence from early Syrian and Turkish plaques that nose-rings and lip rings were used as early methods of control. The practice of using such devices in horses survived for several centuries, but died out probably because of the relative effectiveness and utility of the mouth bit. Because of the sensitivity of the horse's nose, the use of a nose ring must have made horses very reactive to any pressure on the rein, and probably very head shy. Until habituation occurred, this would have made control in all but the quietest of horses difficult. It is surprising, then, that this technique was used for several centuries but this may reflect the tendency of humans to use techniques traditional to their culture without evaluating how successful they were.

Metal bits developed from twisted bronze bars in the fourteenth century BC to jointed bits with the familiar 'nutcracker' action of the modern jointed snaffle in the tenth century BC. Over the next five hundred years, as the uses of the horse diversified in different cultures, bit designs proliferated to give rise to the range of types that are seen today. Each bit has a slightly different action by causing pressure on a different part of the horse's mouth or head, and hence motivating the horse to relieve the pressure through slightly different actions.

Among the most innovative and therefore successful early equestrian cultures, the Scythians developed novel methods of hobbling, such as by roping a fore leg to the diagonally opposite hind leg, and the uraga, a rawhide lasso looped onto a long pole to aid capture. Their bridles, remarkably similar to those used today, were made of horse hide, and were kept supple by being rubbed with animal fat. Their basic saddles were made of leather cushions stuffed with hair, with felt pads underneath, wooden bows at each end, and an overgirth (Richardson 1998).

The Tartars caught wild horses by riding them down and lassoing them around the neck or leg (Berrenger 1771). The horse was hobbled, restrained by neck ropes, and then saddled, bridled and mounted. It was ridden until it stopped all attempts to escape or unseat the rider. This technique relies on the horse learning that whatever behaviour it uses to escape from the restraint is ineffective – learned helplessness (Lieberman 1993). In order to ensure that the horse was compliant in all circumstances, the process described above was repeated at each stage of 'training' so that the horse learnt that it was fruitless to resist restraint in any situation. The horse was ridden surrounded by more experienced horses every few days until accustomed to stimuli from the rider. The newly broken and fearful horse naturally wanted to stay near its conspecifics, and the tribesmen used this tendency to train rudimentary responses. For example, urging the horse on and stopping were associated with the use of the whip or leather thong and bit respectively, at the same time as the rest of the group performed the appropriate behaviour. The horse was then most likely to associate the appropriate response with either stimulus.

The Tartar herdsmen recognised that gradual habituation of young-stock enhanced the level of training achieved later. The steppe ponies were quite small and slow developing and hence the men did not ride them until they are five or six years of age. Foals born in the herd were trained slowly, with the children starting to handle and ride them from six to eight months of age, gradually habituating them to halters, and handling them for progressively longer periods of time. This provided a long period of gradual habituation to handling, saddling, having a mounted rider, and moving away from their herd and made the process of breaking them at five or six years of age easy in comparison to the wild-caught stock.

Horses were introduced to the Greek Empire by the eleventh century. The Persians had developed the use of the horse and chariot as an effective weapon of war against the Greeks, and this was a powerful impetus for the Greeks to develop a cavalry themselves. Effective use of the horse was often a pivotal factor in the outcome of conflicts at this time (Anderson 1961). Largely because of their ability to breed, train and use horses effectively by the sixth century BC, the Persians were a dominant military power in the east. The traditional methods of breaking that had been passed down from the Steppes tribesmen and the Persian Empire were re-evaluated by the Greeks. We can assume that their methods of training were probably not based entirely on kindness because they not only strongly advocated the use of hobbles, but also advised that horses in training should be muzzled to prevent them from biting. Like the Persians they used professional horse breakers, who caught and broke the horses, and performed gelding and other veterinary procedures. In his writings, Xenophon (translated by Morgan 1979) recommended the use of professional horse trainers for young cavalry horses. He advised that foals should experience kind handling, before being trained, and that owners should keep a close eye on how their mount was trained. In addition, he suggested that young horses should be led through crowds and made familiar with all sorts of sights and sounds. This would serve to habituate them to busy environments and prepare them for battle situations later in life. Xenophon also postulated that during the process of horse training, the horse should associate being alone with being hungry, thirsty or annoyed by flies, and the presence of a man with food, water and relief from the flies. This was the first sign of horsemen developing a philosophy that used the horse's motivation to influence the way it behaved. He also went on to write that a horse should be rewarded with kindness for doing something right, although he also said that it should be punished for doing anything wrong. The Greeks used the voice a lot to calm or rouse horses during training, and by classical conditioning, trained them to associate different words or noises with appropriate operant behaviours (Anderson 1961). Interestingly the use of the voice as an aid to training is less frequently advocated in modern training journals, indeed riders are penalised for use of their voice in dressage tests.

Horses to be used in harness were broken from the age of two using long reins. Riding horses were probably also broken at about the same age, although it seems that they were not considered to be ready for use by the cavalry until the age of six (Morgan 1979). Bits with 'keys' attached were also introduced by the Greeks – with the aim of encouraging the horse to salivate and to chew, while moving the bit around in the mouth. This helped to ensure that the horse had a moist or wet mouth, which is considered to be a sign that the horse is 'accepting' the action of the bit. Starting bits with 'keys' attached are still often advocated for this purpose today (see Figure 1) For adult horses the Greeks generally preferred to use bits with a mild action, although young horses were thought to need a stronger bit for a while to teach them to 'respect' the bit. The Greek trainers realised that light pressure on the bit was effective only if it was not present all of the time. Heavy hands caused the horse to habituate to a level of discomfort and to ignore subtle rein signals. The reasoning appears to be that subsequent use of a milder bit would be predictive for the horse of the pain it had previously experienced, and hence only light pressure would be required. A sensitive mouth was considered important, and hence riders and handlers refrained from pulling on the bit unless it was necessary. To preserve sensitivity in their mouths, horses when being led, had a lead rope attached to a halter or the noseband, and riders were taught to ride with 'light' hands. The reins were held by being passed over the forefinger and through the hands to encourage this.

The modern conventions of bridling, girthing and mounting from the left side were also developed by the Greeks. Because horses were probably handled by a number of different people – slaves, trainers and their aristocratic owners, a standard procedure meant that the horses had a predictable routine, and learnt to react to

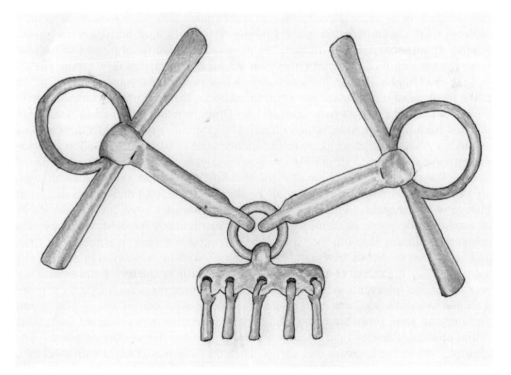

Figure 1. A starting bit with 'keys' attached.

signals given in a specific sequence. The Greeks also considered it important that when bridled, the horse should be led from both sides to prevent it from developing preferences for one particular side. This also prevented habituation to pressure from the bit on one side (Anderson 1961).

Young horses were initially 'backed' (*i.e.* a human sat upon their back) without a saddle. However as their training progressed, a saddle cloth without stirrups was used and finally a full saddle was placed on their backs. Standing still to be mounted was considered an important part of training, and some horses were taught to stretch out their front legs, bend their knees or drop one shoulder to help the rider mount (Anderson 1961).

Interestingly, the Greeks developed a method of breaking their horses in small, enclosed circular arenas filled with deep sand or shavings. These rings had high, solid sides so that the horses' attention was not distracted from their handler. Their circular shape enabled the handlers to keep the horses moving during training. In essence, training in such rings represented another form of negative reinforcement (see following sections). The horses learnt that they could not escape from their trainer by moving away and also that, because the ring was small with deep sand, it was futile for them to keep trying to escape. Hence, they rapidly learnt not to move away from the trainer when in the ring. This enabled the trainer to approach the horse and gradually habituate it to being handled and restrained.

Roman horses were kept in stalls, loose boxes or loose yards on bedding of powdered dung or straw. Because horses were ridden extensively on the widening network of roads constructed in the Empire, those with hard hooves were prized. Leather shoes, sometimes with metal soles, were frequently employed to prevent horses becoming footsore, particularly on long journeys or military campaigns.

Like the Greeks, Roman horse breeders advocated regular handling of foals, to make them more tractable on weaning. In addition, they commonly left foals with their mothers for the first two years of life. Draught mares returned to work with the foal running along side, which effectively habituated the foals to the kinds of events they would be most likely to experience later in life. Mares used to produce riding horses were bred from only every second year (Hyland 1990).

The Romans started the process of breaking at two to three years of age. However, they recommended that foals be stabled and started on a cereal diet from the age of five months. Items of equipment, such as bridles, were hung in the stable to accustom the horse to them, and the foal was regularly handled. Breaking was achieved gradually, as with the Greeks, by having a boy lean over the horse first, and then sit on the horse. Circular pens or 'gyrus', such as those used by the Greeks were used for breaking. Horses were given powerful purgatives before breaking was started, an intervention that would have dehydrated them and made them more tractable to handling. These early stages of training were carried out by slaves and grooms, and were probably recommended more because they produced successful riding or chariot horses than because of any regard for the welfare of horses. For example, we are now aware that abrupt weaning in horses is highly stressful (see Chapter 4 for a discussion of this) and advocate that the foal should remain with its dam for a longer period. The Romans, however, suggested this because they found that their youngstock grew larger and faster if they still had access to their mothers' milk. By modern standards the Romans were generally very cruel in their attitude towards their animals. Most Roman bits were harsh, with twisted metal (see Figure 2) and high ports (the bit was designed so that there was a half-moon shape that was raised towards the roof of the horses' mouth). Some bits even required teeth to be removed before they could fit in the mouth (Hyland 1990). Some bits had metal cheek pieces extending down to join up under the chin forming an early type of curb. They commonly incorporated a metal ring structure that ran from the bit around the front of the nostrils, which not only put huge pressure on the front of the nose, but also made the action of the port more severe. In addition, since the Romans preferred their horses to pace rather than trot, they hobbled the horses' fore and hind legs on each side together to prevent the legs moving in diagonal pairs, as they would do if trotting naturally. This technique was employed in Europe right through to the sixteenth century, appearing for example in articles on horsemanship (Morgan 1609). In order to fulfil their fashion for high stepping horses, the Romans also tied pieces of wood to the pastern joints of their horses to encourage them to lift up their feet. This, again, was still prevalent in sixteenth century England. In a bid to keep them manageable, horses were also generally worked extremely hard as soon as they were broken.

With the diversification of the use of the horse for riding, driving, racing or circus, distinct methods of training, and types of trainers developed for each type of activity. Trainers were dispersed right across the Roman Empire, and even though the size of the Empire declined the skills of breaking and riding horses remained in the new developing cultures of Europe and Asia. During this time, there was a considerable movement of cultural ideas across Europe, Africa and Asia. Slitting of the horses' nostrils, for example, was a practice first carried out by the Persians and Egyptians. The premise seems to have been that since equids are unable to breathe through their mouths, slitting their nostrils enabled them to take in more air and work faster or for longer. The Moors in Northern Africa continued this practice, particularly in their racing horses, and occasionally still do it today

Figure 2. Early Roman bit designs resemble the bits still used today.

(Littauer 1969). The South American Indian tribes had little knowledge of horses, but when the Spanish reached the tribes of Mexico in the 1700's, horse culture soon developed. As horses rapidly moved northwards into the great plains of North America, the practice of nostril slitting re-emerged in the cultures of the North American plains Indians for their racing and war ponies (Roe 1955). In addition, it appeared in the Hungarian military in the eighteenth century in the erroneous belief that it would prevent horses from vocalising and ruining surprise attacks. Our understanding of equine nasal anatomy now makes it clear that slitting nostrils would not have significantly increased air intake into these horses, and so this practice is an useful example of a 'fashionable' practice that spread through different cultures despite being ineffective.

Although the majority of cultures in the Ancient world used a variety of gadgets to control their horses, the cultures originating in North Africa were renowned for riding their horses with no saddle cloth or bridle. They controlled their horses with a guide stick that was tapped on the side of the horse's neck to indicate a change in direction, or on the nose to stop. These tribes people were often hired as mercenaries both by the Romans and the Carthaginians for military campaigns, including Hannibal's attack on the Romans in 218 BC. These people initially trained their horses using bridles, teaching them to the stick signals only when obedient to the reins. The two signals were used together until the horse reliably responded to the stick alone, at which point the bridle was dispensed with. The tradition of riding only with a stick has survived in the donkey riders of Northern Africa who can still be seen guiding their bridleless donkeys just by tapping their necks (Clutton-Brock 1992).

Accounts of Arab horse breaking in the medieval period suggested that horses were gradually habituated to handling, restraint, tack and riders from an early age. Foals were weaned at three to four months of age, and reared in close proximity to people, sometimes inside their tents. This practice was dramatically different to that employed by the Romans, and was purported to improve the horses' trust of people (Hyland 1990). Breaking was started at two or three years of age. Severe bits were generally used for control. Alternatively studded nosebands were developed to press into the horse's face when the rein was pulled. These eventually developed into hackamores (bitless bridles) similar to those used today. Like the bits, they use negative reinforcement – the horse learning to turn its head to avoid the pressure on its face.

In medieval Northern Europe, large, heavy horses prevailed as the most sought-after riding animals because they were able to carry armoured knights into battle (Hyland 1994). Cold-blooded breeds, such as the Freisian and Flemish horses were generally quiet to handle, and due to the style of fighting, were not required to be highly mobile or responsive. The skill needed to break these horses was considered not as great as for the eastern breeds. This led to a decline in the horsemanship skills in Britain and Northern Europe. With the development of gunpowder the huge cavalries of the middle ages began to disappear. This was associated with a further reduction in knowledge of horsemanship in Britain. Horses were broken with an escalation of force and cruelty, and use of increasingly severe curb bits, spurs and whips. The attitude of horse breakers was to dominate and subjugate the horse, and keep it obedient through cruelty and sheer hard work. The manuscript on horsemanship written by Rusius in 1486, which advocated extremely harsh methods for managing horses, was an accepted manual of instruction for almost two hundred years. He advocated practices such as beating with whips wrapped in hedgehog skins, and hot iron bars put under a horse's tail if it refused to move forward. Almost every recommended practice during that period involved either punishment or negative reinforcement.

Luckily for horses, the renaissance brought a new philosophy to horsemanship in Europe, which has influenced methods of riding and training up to the present day. Equitation became an art form (Wynmalen 1938), and school riding was popular with much of the European aristocracy. This new school required riders and trainers to have a greater understanding of how their signals and commands were used. Unfortunately, this did not automatically mean that trainers used kind methods of training, or understood the nature of their horses, and it was some time before more sympathetic methods of training were incorporated into this new art form. One of the early methods of training in this new school, probably derived from Moorish practices of tying horses to a post, was to have a post in the middle of the school with a groove cut near the top, to which the horse was tied. The trainer, with a whip, encouraged the horse to move forward in a circle around the central pole. This early practice sometimes involved tying up one of the horse's legs to tire it out more rapidly, if it would not behave appropriately. This practice was also used by American horse trainers in the 1700's, who lunged difficult horses around a pole with one leg tied up and weights on their backs. As with the earlier techniques of training, these methods led to the horses reaching a state of learned helplessness, and giving up all attempts to resist in that context.

In Eighteenth century Europe, lungeing without the post developed into the form that is commonly used today, with the trainer holding both the lunge line and the whip. Although the technique is similar to the pole method, lungeing is now rarely used to create a situation of learned helplessness, but is used to enable trainers to facilitate operant learning in their horses. For example, horses are encouraged forward at the same time that a vocal command is given, so that ultimately they associate the command with forward movement without the need for other stimuli. In addition, lungeing incorporates specific exercises to increase flexibility and muscle development.

The next device to be developed in the Renaissance period was the 'dumb jockey', a construction of two rings or bars attached either side of a girth strap, from which two reins were attached to the bit. This device was left on the horse whilst in the stall or the field, effectively with its head strapped down, to encourage a 'good' mouth and head carriage. Rather like modern draw reins, the device probably just encouraged the horse to lean on the bit, something that is no longer. As is still the case today, devices such as these are popular with those who want an quick and easy method of ensuring that their horse looks as if it working properly, even if it is not.

Since Roman times long reining (see Figure 3) has been an established technique

Figure 3. Long-reining: A traditional training technique that is still used.

to encourage young horses to accept the bit. Devices designed to simulate this technique but with less work on the part of the trainer emerged in Europe in the 1900's. They included reins being passed from the horse's mouth via a pulley on the ceiling of the stable to the trainers' hands. Another method, still sometimes used to train carriage or driving horses, was the use of training pillars. The young horse was tied with ropes, or 'pillar reins', running from its bit to two posts on either side of it (Richardson 1985). This was sometimes done to horses as they stood in their stalls, and is still used by the so-called 'higher' schools of equitation, such as the Spanish Riding School in Vienna.

The use of severe bits was still common in the renaissance period. Forward movement was inhibited by oral discomfort to the extent that whips and spurs were commonly used to get the horses to move forward. The head carriage of the horses was often very low, with their chins tucked into their chests (or over-bent) as they tried to evade the restriction of the bits. These techniques of riding and training continued through until the 1900's. Suggested 'cures' for difficult horses during this period involved some form of beating with various designs of whip. Enlightenment began with masters of the Italian and French schools of equitation who assumed that horses resisted training due to misunderstanding rather than an inherent viciousness that needed subduing by force. This change in thinking was an enormous step forward from the anthropocentric attitudes of that time. The first Briton to publish this approach was William Cavendish (1743). He suggested that horses should be trained gradually over a number of years through a process of gradual habituation – as the Greeks had done. With the development of a type of riding activity called 'dressage' came the use of more subtle aids, such as applied by the leg alone rather than via whips or spurs. Again this is a form of negative reinforcement – the horse learns that the rider stops applying leg pressure to its flank when it has either increased pace as required or moved 'off the leg', for example, to execute a lateral movement.

The huge loss of horses in wars such as the Crimea and Boer meant that the British army had to buy and train horses at a faster rate. Because of the constraints of time, the methods of breaking developed by the army horsemen were quick and by modern standards, often cruel. In general the practices relied heavily on learned helplessness. For example, one process involved the horse being thrown with ropes to convince it of the 'superiority' of its human handlers. The horses were then backed in small pens where riders stayed on until the horses gave up any attempts to rid themselves of the riders. For difficult horses, this process was done with the head tied to the tail or girth, or one foreleg tied up. The horses also wore tight side reins to limit their ability to move. Similar practices were adapted in North America, particularly at the time of the American Civil War, where, again, the demand for cavalry mounts was enormous (Denhardt 1947).

In North America, the Indian tribes of the plains comprised many highly proficient horsemen, and developed handling and riding techniques that were in many ways similar to the ancient Scythians of Asia. A wild horse, once lassoed, was held by several men with ropes until the horse realised that it could not escape. At this point, one man, the trainer, gently approached the horse whilst it was held by the ropes, and repeatedly approached and retreated from the animal whilst speaking to it gently. Once the horse had learnt to tolerate this, the trainer approached, waving a blanket or buffalo skin, until this too was accepted by the horse. The man then approached with a halter made of thin rawhide and put this on the horse's head, enabling him to control the horse through pressure on the nose and poll. From that point on he would continue training the horse alone. He gradually touched the horse all over its head, body and legs, and finally progressed to putting some weight on the back, and then mounted. It is reported that during this process very few horses resisted at the mounting stage, or bucked when ridden (Roe 1955).

The North and South American settlers also used traditional cultural methods to tame members of the growing herds of wild mustangs. In Mexico, horse breakers used methods similar to those of the Spanish conquistadors who first introduced horses to the Americas. The methods involved the subjugation of the horse, using ropes and halters to restrict its attempts to escape. Further South, in Argentina and Paraguay, training methods involved roping the horses, tying them to a post, saddling and then mounting and riding them forward with spurs to the point of exhaustion. Similar methods are employed by South American gauchos today. In North America the same philosophy was adopted, although the methods sometimes varied. For example, objects were tied to the tails of horses in corrals to frighten them until they were too exhausted to run any further. Some ranches, particularly those with more valuable animals, were more careful in their breaking methods, and used a circular pen similar to the Roman gyrus to keep the concentration of the horse on the trainer. These horses were often trained first with 'dummy riders' such as sacks of grain, before real riders mounted. Even after the initial breaking process, methods of further training were frequently harsh. Training a horse to 'neck rein' for example, sometimes involved beating the horse on each side of its neck so that it was sore and swollen and therefore responsive to the light touch of a rein on the next day. The visually impressive skills of the cowboys in taming the wild mustangs led to the development of the rodeo in the 1800's, which became a highly popular form of entertainment. The popularity of these shows, is probably one reason why the traditional methods used by cowboys when training horses have survived so long.

3. Learning in Horses

It is clear from the history of the techniques used to tame horses that modern practices are largely based on traditional methods that have developed since the earliest days of horse domestication. More recently there has been an interest in the application of the scientific principles of learning theory to animal training. Learning has been defined as the modification of behaviour in the light of experience (Thorpe 1963). In theory there are several ways in which humans and animals learn. Increased awareness of the ways in which horses learn and methods for improving learning could greatly enhance the effectiveness of training.

4. Types of Learning

For clarity, the terms used in learning theory as well as some examples of these forms of learning in horses, are explained in this section. For a more detailed explanation of each of these types of learning see Pearce (1997).

- *Habituation and Sensitisation:* These are seen as being the opposite of each other. Habituation is a waning response to a stimulus whereas sensitisation is an increasing response to a stimulus. It makes sense in certain situations for the horse to stop reacting when the stimulus occurs frequently if it is unreliable at predicting anything significant, whereas it is also important for the horse to be able to react quickly to a stimulus that reliably predicts a situation that is biologically significant.
* *Classical conditioning:* This is seen where an already established behaviour is associated with a new stimulus or range of stimuli. An example of this is seen in the learned appetitive behaviours of stabled horses. When they are waiting to be fed their concentrated food in the relative confinement of a stable or loose-box, they will move around and sometimes bang the door with one of their fore-legs whilst waiting for the food. These demonstrations of arousal indicate that horses have learned that the arrival of food is reliably associated with the sound of buckets being moved or filled. Another example is seen in stallions trained for semen collection using a dummy mount. In associating entry into the service barn with the practice of semen collection, the stallion's penis will often become erect without the stimulation of an oestrous mare.
- *Operant conditioning:* This form of associative learning is seen when the horse increases the performance of a specific behavioural response due to the experi-

ence of a positive or negative reinforcer. To ensure that a specific behavioural response occurs in future instances involving the same cues, behaviour must be reinforced. Positive and negative reinforcers can both be viewed as rewards in that they both lead to an increase in the performance of a behaviour. Positive reinforcement is where the desired behavioural response is associated with access to resources the horse would normally desire such as sweet food, whereas negative involves the application of an unpleasant (painful/fear provoking) stimulus until the animal performs the correct behaviour, at which time the stimulus is removed. This results in learning to respond differently. Although reinforcement can be negative, this is not the same as the concept of punishment. Negative reinforcement differs from punishment in that it enables the horse to control its experience. The horse can avoid unpleasant circumstances by performing a specific response. For example, a slight pressure with the inside of the rider's leg will signal to the horse to go forward, and as soon as it moves, the pressure is taken off.

Punishment (the deliverance of an action that stops a behaviour or reduces the incidence of a behaviour) occurs after the response. The horse then has to make an association between what occurred in the past and the painful or fear-provoking circumstances. Since it involves retrospection, the horse has no control over the situation. An example of this may arise when a horse jumps over an obstacle and unbalances the rider. In this situation the horse is punished for jumping well over the fence, because it receives a painful jolt in the mouth and back as the rider struggles to stay with the horse. If this occurs regularly or indeed intensely on just a few occasions, the horse may start to fear jumping. Punishment is an inappropriate method for dealing with problem behaviour in horses. Although it is often the method that is first tried by owners, it has many problems. In order to be effective it must be delivered very quickly (*i.e.* within a second or two) after the occurrence of the undesirable behaviour and even then there is a danger that the horse will associate the punishment, with the wrong aspect of the situation. For example, many riders deliver punishment to a horse if it refuses to jump over an obstacle. Punishment normally involves flagellation of the flanks and is delivered once the rider has gained control over the horse, which may be a few moments after the refusal. Unfortunately, such punishment may be associated with 'settling down' after the evasion, and therefore does nothing to show the horse what it should be doing. It may also serve to frighten the horse. When an animal is fearful, it will find it difficult to learn any new responses, other than those that satisfy its motivation to escape. In addition the horse may learn to fear the rider, or the fence, which could lead to additional problems.

- *Social learning:* This type of learning can occur when one animal observes the behaviour of another. In a social species such as the horse, where the young remains with the dam for an extended period, it is possible that much of what it learns about finding and selecting food as well as its social behaviour will be through observation. Although there is little scientific evidence that horses learn in this way, this is probably because the methods used to study it in the horse have not yet been made biologically relevant. For example, most published studies have used a single demonstrator horse that has learned a task and carries it out while test horses observe (*e.g.* Lindberg *et al.*, 1999). The observer horses are then studied to measure their ability to perform the task itself. So far no good evidence of observational learning has emerged using this paradigm. The variable relationship between the demonstrator and observers may account for this and it may be that to act as an effective tutor, the demonstrator horse needs to be closely related or at least well known, to the observer horse.

5. Constraints on Learning

There are biological constraints on what an animal can learn. Animals are predisposed to learn certain responses during defined stages in their development. Although some owners will claim that their horses think as they do, there is no evidence in the horse, of the so-called 'higher learning' capabilities, such as seriation, that are seen in humans and chimpanzees. However problems arise for the horse where the handler or trainer expects the horse to make abstract associations. Such anthropomorphism leads to situations where any relationship between horse and owner/handler is likely to break down. Leaving a horse tied up to a fence to 'think about' how bad it has been, is inhumane and ineffective. Any associations the horse makes will most likely be unhelpful. The effects of this so called 'hardwiring' of behaviour have been eloquently described by Breland and Breland (1961) in their work on the misbehaviour of organisms.

Horse behaviour changes are associated with increasing maturity and changes in experience and situation. To understand what motivates an animal to perform a behaviour it is necessary to understand the relationship between the various internal stimuli that are involved in the control of the behaviour and the external factors that elicit the behaviour. Because of physical growth and changes in neural organisation, changes in motivation can affect learning during developmental periods in the animals' life. For example, an entire male will be less motivated to learn a behavioural response that helps him access a female if he is sexually immature (because his internal state differs from a mature male), or if the female is not in oestrous (in which case the external stimulus is less attractive).

5.1. SENSITIVE PERIODS

Animals are more likely to learn about certain aspects of their environment at particular stages in their development. These sensitive periods have been likened to 'windows' in development. For example the newborn foal will form an extremely strong following attachment to the first large, moving object with which it has contact (Tyler 1972). Under normal conditions this ensures that it 'bonds' with its mother. This opportunity for learning on the foal's part coincides with maternal bonding, during which the dam learns to discriminate between stimuli (smell, sound and sight) that identify her new-born.

5.2. PHYSICAL GROWTH

Certain behaviour is possible only when the horse is of a particular size or strength. For example, when learning to forage, horses may feed on tough grasses only when they are old enough to prehend and masticate such material.

5.3. NEURAL ORGANISATION AND MATURATION

Certain associations may not be learned until the animal is at a given stage in its neural development. This may be due to the animal lacking the sophisticated co-ordination to perform the action. In addition, it may not be able to remember the response or, for that matter, the outcome. Fraser (1992) points out that long term memory is important for successful learning. Particular neurotransmitters (*e.g.* acetylcholine) are implicated in the learning process and it appears that sensitivity to neurotransmission increases with learning experience (Prosser 1991).

6. Applying Learning Theory: Traditional Training Practices

As outlined earlier in this chapter, it has been traditional practice to change a horse's behaviour through negative reinforcement, whereby an unpleasant stimulus is applied to the horse until it exhibits a desired response. This approach to training horses is still very much in evidence today. For example, depending on the bit it incorporates, the bridle is designed to apply pressure, discomfort or pain in one or more of five sensitive areas. These are the tongue, the bars of the mouth, the chin, the palate and the poll.

Modern day horse trainers talk about training the horse to 'find the comfort zone'. This describes the way the horse learns that by putting his head and neck in a certain position it can reduce the pressure from the rider's hands through the bit. The reward for bringing his head down and relaxing his jaw (and thus accepting rather than fighting the action of the bit) is that he will find relief from the discomfort of the bit action. Thus the motivation of the horse in this instance is to avoid stimuli that alert him to potential harm. By the same token, leg pressure motivates the horse to move forward or laterally away 'off the leg'.

Positive reinforcement is another method for training. A horse is more likely to perform a behaviour that is associated with a pleasant experience. Care should be taken however that the number of repetitions the horse has to make before a reward is given are not too numerous, in which case the association becomes extinct. This is when the signal the horse uses to predict the delivery of a food reward is no longer reliable. While food is commonly used as a positive reinforcer in most horses, access to social companions can also be a reward for horses, because they are highly dependent on conspecifics for safety. The relevance of the reinforcer is important. For example, when training a horse to load into a horse box or trailer, an appropriate positive reinforcement might be access to a companion. However, the identity of the companion is important; if it is a close social companion then motivation to load may be greater than if it is an unfamiliar horse. Indeed some conspecifics may actually deter a horse from loading either because the naïve horse is relatively deferent or fearful.

The reinforcement schedule describes the way in which the reinforcement is applied. In the first instance, when the association is being established, the horse needs to be rewarded every time he performs the desired response. In this way the horse learns the amount of effort required to get a predictable reward. Once the association is established, a variable reinforcement schedule is most likely to keep the horse performing. Under a variable reinforcement schedule, the horse is not rewarded every time he performs the behaviour, and therefore the reward is unpredictable. A variable reinforcement schedule motivates the horse to perform the behavioural response every time a particular signal is given, and it allows the trainer to reward only those efforts that are particularly good, and so increases the accuracy of the response. This is called shaping.

Once the desired behavioural response has been linked with the food reward or the primary reinforcer, the association can be further refined. Performance horses need to perform accurate and sophisticated movements in many different situations. This requires that the horse learns to generalise certain associations so that the behavioural response can be performed in any situation. This is important, since it is extremely easy for learning to become context specific. In addition the horse must be able to perform the behavioural response in situations where it is not possible to reward the use a primary reinforcer.

Horse training can involve the use of secondary reinforcers (sometimes called bridges or bridging signals). A secondary reinforcer is a stimulus that the horse learns to associate with innately valuable resources. Secondary reinforcement might involve the use of a word such as 'good' or a 'click' of clicker, or a pat on the neck. The secondary reinforcer lets the horse know that it is about to get the desired reward and so bridges the gap between the performance of the correct behavioural response and the desired consequences.

The last stage of a training process is usually to add a signal or 'aid' to the association, so that the animal performs the behaviour in response to that specific signal. For example, standing still, may be linked with the word 'stand' or the action of raising a hand. If the specific signal is applied to the behaviour after it has been trained, then the horse is unlikely to link the signal with inappropriate responses (such as those that it performed during training). For example, if the word 'stand' had been used whilst training the behaviour, the horse may have learned to associate 'stand' with the need to struggle against, or at least anticipate, pressure being applied through the head-collar.

6.1. RULES OF SHAPING A BEHAVIOUR

Many of the horse behavioural responses required by riders are achieved through a process called shaping. Shaping is the process by which a new behavioural response is established through a series of stages, or successive approximations towards an end goal. The process of shaping involves the following steps:

1. Break down the desired manoeuvre into a number of behavioural responses.
2. Work on one response at a time, ensuring that each step is within the capabilities of the horse and can be performed within the limits of the situation and time available.
3. When moving on to a new element, relax the standards on the previous ones.
4. If at any time the horse appears to be having problems with the task, take him back to a previous stage of training.

7. Problems with Applying Learning Theory

Clarity, consistency and kindness are the essential ingredients for successful training. Sadly, training tends to be thought of as a confrontational process and is often based upon 'dominating' the horse (see Cregier 1987). Dressage scoring is based on a number of features, including how much 'submission' the horse exhibits. Additionally many training manuals advocate that if the horse does not respond to a signal, it must be reprimanded with the whip, or have the aid constantly repeated (*e.g.* Marshall 1989). The use of fear and pain in training may contribute to later problems in the horse's working career, as the horse may learn ways of evading a fear-provoking situation and sometimes become unmanageable.

Although perceived as behavioural problems, evasive responses such as frequent bucking or rearing when experiencing pain (from inappropriate handling, riding or ill-fitting equipment), are the natural equine responses to threatening situations. Horses performing these responses are often labelled as rebellious and are sometimes punished, an approach that often leads to an escalation of the problem. Poor training and handling leads to problem horses whose value and welfare are compromised.

8. Factors That Affect Training Success

The trainability of the horse describes the ease with which it can be trained to respond appropriately to various signals commonly used by riders/handlers. Recently, considerable attention has been drawn to the work of horse trainers who seek to work with, rather than against the natural behavioural traits of the horse (*e.g.* 'Imprint training', Miller 1989; 'Join-up', Roberts 1997). These highly skilled horse-trainers have raised awareness of the importance of understanding equine behaviour, even though their methods are often not based upon scientific studies.

8.1. LEARNING ABILITY

Empirical studies of the perceptive abilities of domestic horses offer a means for understanding the way in which horses learn and provide practical information for aiding the training process (see Chapter 1) (Flannery 1997). It is obvious from such studies that horses vary widely in their ability to learn, and that opportunities to

learn early in life assist in the training process (Heird *et al.*, 1981). Highly emotional horses, those that show the greatest fear responses to novel stimuli, score less
well in learning trials (Heird *et al.*, 1986), and there is some evidence that breed
differences also exist. For example, Thoroughbreds are reported to learn less well
than Quarterhorses, (Mader & Price 1980), perhaps due to their early developmental
experiences. Learning ability also appears to decline with age. No relationship has
been found between dominance status or sex in ability to learn. This is despite
anecdotal reports that dominant horses can be difficult to train and are more likely
to find a means of evading the trainer (Budiansky 1997).

8.2. EARLY EXPERIENCES

The behavioural responses horses show are products not only of inherited genes,
but also of the rearing environment. If they are to be expected to learn well, they
should be reared in naturally complex environments that give them the opportunities to develop their learning abilities early in life, when advantage can be taken
of plasticity in their behaviour. Simply turning a young horse out to pasture and
expecting it to develop the behaviours to suit human requirements, is not sufficient.
    Emotionality in many animals tested under laboratory conditions is influenced
by the frequency and manner in which those animals were handled during early
development. For example, socialisation periods for dogs are thought to occur
between 8 and 12 weeks of age, a time when ability to learn social behaviour is
high (Scott & Marston 1950). If handling by humans does not take place at this
time, later trainability is reduced. Despite the obvious value of such comparable
equine data, early handling of horses has received little scientific interest. Handling
throughout the first 42 days of life has been shown to increase foal performance
in a halter-training task compared with handling from 43 to 84 days of age (Mal
& McCall 1996), but there is no evidence that pre-weaning handling has any effect
on the post-weaning learning ability of foals (Mal *et al.*, 1994). The timing of
handling is important perhaps due to differences in the propensity for the development of learned helplessness. Horses that have been extensively handled learned
which way to turn in a maze more slowly than minimally-handled horses, but faster
than un-handled horses, suggesting that only moderate handling is best from the
horse's perspective when in a problem-solving situation without human help (Heird
et al., 1981). It is known that orphan foals reared by humans are less emotional
when placed in a novel environment when compared with normally reared foals
(Houpt & Hintz 1983). This suggests that early handling at a particular time during
development can influence reactivity later in life. Indeed Miller (1989), suggests
the use of 'imprint training' for newly born foals since the earliest learning in foals
is a form of imprinting. Imprinting is probably more important in precocial species
such as equids, in which the offspring are less dependent on their mothers for food
and warmth, than in altricial species, which often confine their more vulnerable,
and often hairless, young to nests. Most imprinting promotes survival of foals and
may shape their future breeding activities. Imprinting has a number of characteristics, including a critical sensitive period.

Despite the apparent relevance and perceived benefits of early handling, horses under commercial conditions are rarely intensively handled until they are considered sufficiently physically developed to carry a rider, at about three years. The racing industry train and mount their horses much earlier (a year old in some cases). Despite some objections concerning the young age at which racehorses are expected to perform, this practice may have welfare benefits because it exposes horses to relevant stimuli while they are relatively impressionable (this is discussed further in Chapter 8).

8.3. TEMPERAMENT

Temperament describes a horse's 'personality', emotionality or reactivity. It is usual amongst the horse owning fraternity, to refer to a horse as having a certain type of temperament (Mills 1998), it is assumed that temperament is inherited (e.g. Williams 1997) and horses are often selected for breeding at least partially on the basis of their temperament. However, there is very little empirical evidence to support this notion, and it is only recently that scientists have become interested in determining whether horse temperament is a measurable property.

The idea of using a simple test early in a horse's life to determine its personality type, or temperament, is attractive and there is potential for such a testing scheme/programme. If tests were reliable predictors, the suitability of a horse for a method of training, a particular discipline, and a certain living environment, could all be assessed early in life. This would allow a matching of horses for specific roles/sports and avoid the wasting of time, energy and money on rearing and training horses that were not suited to a particular task as well as helping to prevent the development of unwelcome behaviours.

The temperament tests that have been carried out have looked at horses' responses to being placed in an unfamiliar arena, being presented with an unfamiliar or novel objects, people and obstacles. It is clear from these tests that horses differ in their initial responses to novel stimuli (see Seaman et al., 2002, for example). However, most of the tests have failed to show that this difference is sustained over a number of similar tests. Furthermore they tend to measure horses' responses to the environmental challenges with which they are broadly familiar. To be useful, tests of learning ability must relate in some way to the horse's trainability, and tests of fearfulness should relate to the horse's ability to cope with new challenges (see Visser et al., in press).

Interestingly, horse-riders speak freely about temperament and they assume that others understand what is meant by subjective terms such as, 'excitable', 'laid-back', 'stubborn' and others. However recently a study testing the reliability of these labels (Mills 1998), showed that when a number of experienced horse trainers were asked to describe the same horses, they did not agree with each other's assessments. The only labels that were reliably used to describe the same horses were 'flighty' and 'sharp'. These results are of concern since some horses could be 'labelled' by one trainer/rider early on in life, and any consequent training and handling techniques may be influenced by this assessment. However, Visser et al.

(in press) recently reported good agreement among a panel of naïve riders when assessing riding school horses, although the temperament traits described by the riders seemed to relate to only two main components; 'responsiveness to the environment' and 'attention to the rider'.

It seems likely that there are core 'personality' traits that might be constant and that can be used to predict the way a horse might respond to novel stimuli and different management and training regimes (see Seaman *et al.*, 2002). More research concerning the objective assessment of individual variation in the behavioural responses of horses to different situations is required. Only then will it be possible to develop a reliable suite of equine temperament tests.

9. Applying Learning Theory: Dealing With The Problem Horse

Various methods are used by trainers and horse behaviour specialists when attempting to modify a horse's behavioural response. Most are based on the assumption that the problem behaviour has been learned, and is therefore rewarding in some way. Although in some cases aversion therapy is necessary for a horse to learn the negative consequences of its behaviour, this can still be carried out in a humane way, especially if the horse is given opportunities to learn undesirable consequences of unwelcome responses in a controlled and systematic way.

Many behaviour problems arise because a horse has been rewarded for its inappropriate actions, without the owner/rider realising. Horses are extremely good at learning how to avoid potentially aversive situations. Take, for example, a young horse that is highly motivated to stay with its group mates. Its natural response when attempts are made to make it walk away from them, is to resist. If the resistance is successful, and the horse gets to return to its group-mates, it will try the same tactic again in the future. Having been reinforced, the horse may even work harder to regain contact, and the rider or handler may be forced to give in again. So-called 'napping' (where the horse resists attempts to make it leave the company of other horses) can therefore become a learned response to being taken away from other horses. The way that problems such as this are avoided, is to ensure that training programmes are designed for each horse, and the horse is not put into a situation where it can learn the wrong associations. A greater appreciation of the importance of all experiences the individual horse has had during development helps to ensure successful training.

10. Techniques Frequently Used in Behaviour Modification

Counter-conditioning is useful when the horse has already acquired an unwelcome behavioural response to a specific stimulus. Its aim is to teach the horse a more appropriate response. Such substitution of one behaviour for another, can be achieved by ensuring that the horse is taught the association between a new behaviour and the reward in a neutral situation (*i.e.* one that has nothing to do with the context in which the problem behaviour occurs). For this to work the horse must be highly motivated to obtain the reward. An example of the use of this technique is when a horse behaves aggressively when given its concentrated feed in a bucket. A way of treating this is to change the place in which it is fed, and to teach the horse to stand quietly to be groomed whilst being fed by hand. Gradually the food can be introduced in a bucket, and the horse will accept the presence of a handler whilst feeding. Once the new behavioural response to food has been established in this way, the horse can then be returned to the original context (Voith 1986).

Another commonly used method is systematic desensitisation. Desensitisation is a humane method of reducing fear towards a stimulus. This is often used to help a horse overcome so-called phobias or other anxiety based problems. In contrast to other techniques used to overcome fear responses (such as flooding), the horse has control over the situation, since if it shows any anxiety during the process, the trainer returns to an earlier stage. Systematic desensitisation can be achieved by ensuring that the horse is in a relaxed situation whilst it is exposed to gradually increasing representations of the feared stimulus. For example, horses that are fearful of loud noises, can be gradually desensitised by playing a tape-recording of the fear-provoking sounds under controlled conditions. Whilst the horse is in a relaxed state, the recording can be played at low volume; increases in volume only being attempted if the horse remains relaxed.

Sometimes when the horse is exposed to a fearful stimulus until it no longer responds (flooding) there is an erroneous belief that it no longer fears the stimulus. During flooding the animal has no control over the stimulus. This differs from systematic sensitisation because with flooding the horse is exposed to the full frightening experience in a situation from which it cannot escape. This lack of control during flooding can lead to the development of extreme fear and learned helplessness to the point that the horse stops responding to various external stimuli, a state labelled apathy (Fraser & Broom 1990).

11. New Innovative Training Methods for Horses

As previously discussed, the methods used to train horses in different cultures have been related more to the attitude of that culture to animal welfare, ethical principles and necessity, than effectiveness. Training techniques involving coercion and cruelty have been used by individual trainers for 'demonstration' horses, and hence have been propagated to other trainers. What is less obvious is the overall success rate for these practices. Although our understanding of learning in horses is still in its infancy (Clarke et al., 1996), it is well known that fear retards learning in other species (Lieberman 1993). It is especially important to consider the role of aversive stimuli in the training of prey species such as the horse (Voith 1986), which will readily adopt flight responses when exposed to aversive stimuli. Inducing a fear response, therefore, is not only likely to reduce the ability of the horse to learn tasks, but also compromises the safety of riders and handlers (Press et al., 1995). Handling techniques that reduce fear in horses can improve both operator safety and horse welfare (Blackshaw et al., 1983).

Some of the currently popular and apparently novel approaches to breaking and training horses are therefore examined below:

11.1. TRAINING/HANDLING HORSES FROM AN EARLIER AGE

Many horse breeders recognise the life-long benefits of thorough handling of their new-born foals in the first week of life. The dominant sense involved in equine imprinting is probably sight, and visual stimuli that youngsters are exposed to during this period will be accepted as 'normal'. The associations that a foal makes during the imprinting process are irreversible and retained for life.

There are welfare advantages for a foal that learns to tolerate rather than struggle. That is the central purpose of so-called 'imprint training'. Good 'imprint' training is said to reduce the prevalence of defensive aggression (Miller 1998). Because imprinting is virtually indelible, time spent working with foals in this way is thought to be very efficient. Miller advocates ritualised habituation of the foal to common stimuli and then sensitisation to selected performance-related stimuli. In this way he fosters passivity while preserving a degree of responsiveness. He advocates between 30–50 stimulations of each area of the foal's body including ears, including ear canals, face, upper lip, mouth, tongue and nostrils, eyes, neck, thorax, saddle area, legs, feet, rump, tail, perineum and external genitalia. Next he introduces what one might generally consider 'unusual stimuli'. For example, the handler goes on to rub the foal's entire body with a piece of crackling plastic until it evokes no panic response. Desensitisation of the newborn foal to gunfire, loud music, flapping flags and swinging ropes follows.

After habituation, in Miller's recipe, comes sensitisation. He teaches the foal that resistance to pressure applied to the flanks or to the head, via a head-collar is useless. Reward in this part of the programme is given by relieving the pressure. It is possible that by imprinting some foals can learn to be too familiar with humans and even develop a social preference for humans over equids. As with any training, the key to finding the best course is to be consistent. If you don't want a horse that takes dangerous liberties in later life, then never reward it for doing so as a foal. The human handler should do the same to avoid the youngster growing up with a list of inappropriate responses to cues from humans. This explains why sensitisation (teaching the foal to move away on cue) is so important in so-called imprinting programmes.

11.2. USING POSITIVE REINFORCEMENT

Studies in other species indicate that positive reinforcement facilitates learning (Lieberman 1993), because the horse is motivated to perform the behaviour without fear decreasing its ability to learn. However, the benefits of reinforcing desirable behaviour with a primary reinforcer in the horse have yet to be studied comprehensively under scientific conditions. Apart from the benefits of improving the ability of the horse to learn new tasks, the use of reward based training has a positive effect on the relationship between the horse and owner. The use of positive reinforcement may lead to bolder and more 'inventive' equines compared to those that have learnt to avoid trying anything new, for fear of punishment. This is because traditional training techniques based on negative reinforcement, tend to train horses to expect correction as a common sequel to novel responses. It is also likely that methods using positive reinforcement produce fewer 'problem' or dangerous horses that learn undesirable behaviours to escape apparently unavoidable punishment.

One of the major problems, and criticisms, of using positive reinforcement, is that it is difficult to administer in practice. It is really only feasible to use food as a reward, because other primary reinforcers, such as access to conspecifics are less practical. Some handlers worry that using food rewards in training a horse from the ground will cause some horses to learn to nip or 'mug' handlers for food treats. This is avoidable if the animal is never fed on demand. Primary positive reinforcement may be less feasible in equitation, because the rider has to slow down the pace and reach forward to the horse's mouth with a piece of food. Predictably this tends to make the horse slow down and even begin to turn around to look at the rider, in anticipation of a reward. Thus the horse associates the reward simply with a reduction in pace and change in neck posture rather than the behaviour that was being targeted for a reward. This drawback does not apply when bridging stimuli are employed. To overcome the problem of reward delivery, a simple rider-operated pump that delivers a small food reward (carrot juice) directly into the horse's mouth during athletic activity, is currently under development (unpublished data, McGreevy). Since horses are more likely to perform a given behaviour if the desired response is *contiguously* associated with a food reward (Lieberman 1993) and this device can instantly deliver food rewards to the horse's mouth this means that the time between performance of the desired behaviour and its reinforcement can be minimised; effectively enhancing the speed of learning (Rubin *et al.*, 1980). Telemetric technology allows the pump- delivery system to be activated by trainers on the ground which facilitates remote intervention during the education of horses being broken-in and schooled either on a lunge-line or loose in a paddock.

11.3. USING CONDITIONED POSITIVE REINFORCEMENT

One commonly used form of conditioned reinforcement is clicker training, which classically conditions an association between the sound of the clicker and a food reward. The association allows the trainer to bridge the gap between the time at which an animal performs a response correctly and the arrival of a primary reinforcer. Secondary reinforcers are most effectively established when presented before or up until the presentation of a primary reinforcer. Critically the clicker sound should *always* be followed by a positive reinforcer. Simultaneous presentation of a reward and a novel secondary stimulus works less well because the primary reinforcer seems to block or overshadow the new stimulus. Similarly, presentation of the secondary stimulus after the primary reinforcer is unhelpful, because although an association may grow between the two, it does not help the animal to predict the arrival of a reward. Using a conditioned reinforcer such as a clicker, then, has three advantages over the use of primary reinforcers directly. The click can be used immediately after an appropriate behaviour, making it easier for the horse to associate the correct behaviour with the reward. Secondly, consistently using a signal to predict a food reward means that the horse will not expect food from the trainer unless it hears the click, which reduces unsolicited requests (or 'mugging') for food in the early stages of training significantly. Thirdly, using a consistently reinforced predictor means that learning should be faster then when a primary reinforcer is used alone, although this has not been tested scientifically.

An advantage of a commercial clicker device is that the sound it makes is distinctive. However, as long as it is not easily confused with words that appear in common parlance, any human vocalisation can be used. Often termed clicker words, these have the advantage of being always available. The use of clicker training principles for shaping and modifying unwanted behaviours, such as reluctance to load, merits serious consideration. By deconstructing a response we can demonstrate to the horse that its fears may have been unjustified. Rebuilding the desired response then becomes a reasonably simple chaining (or sequencing) task. Clicker training is successful in equine behavioural clinical practice, especially in modifying established fear responses.

11.4. THE APPROACH AND RETREAT METHOD

A number of breaking methods have developed over the last century, which have had similarities to those used by the American Plains Indians. These techniques share the principles of:

- isolating the horse from all other distractions so that its attention is focused on the handler;
- using calm movements especially when gradually approaching the horse so that it can habituate to increasingly close contact;
- ensuring that the horse is dependent on the handler for relief from aversive stimuli that arise in the training process.

An Australian trainer called Jeffery developed a method, called the 'approach and retreat method', using these principles in the early 1900's (Blackshaw *et al.*, 1983). His technique, which involved having the horse in a rectangular pen with a loose rope on its neck, was very successful. As the horse moved around the enclosed space the rope tightened as it moved away from the trainer, and slackened when it came nearer. Once the horse had learnt that being nearer the trainer was its best option, it was gradually habituated to progressively closer contact and handling over allogrooming sites culminating in touching the animal in more sensitive areas such as the inguinal region.

11.5. 'SYMPATHETIC HORSEMANSHIP'

In recent years natural horsemanship, gentling and whispering have received tremendous attention from horse-owners. However, the ideas behind these movements are not new. In his book, *Horse Training: Outdoor and High School* of 1931,

Captain Ettienne Beudant wrote about training a horse by seeking his 'approval, his good humour and consequently his obedience', as opposed to the more traditional methods of forced submission.

The fundamental philosophy of approaches that avoid submission is that the horse comes to regard the trainer as the alpha member of a two-member 'herd', rather than a predator. Observations of wild horses have shown visual cues delivered as body language to be the cornerstone of intraspecific communication (McGreevy 1996). For this reason, sympathetic horsemanship techniques emphasise the importance of body language in communication. The philosophy that underpins the approach used by leading horsemen such as Monty Roberts is that to discipline a herd member, for example an unruly colt, the lead mare may exile him from the herd until he shows adequate signs of submission. These signs are believed to include lowering his head to the ground, chewing, and licking his lips. To send the horse away from the herd, the mare will face the miscreant squarely and look him in the eye. She may chase him away if necessary, but usually body language is sufficient to drive him away. When the mare is prepared to let the horse back into the herd, she will take her eye off his, and show him part of her long axis (Roberts 1997). Unfortunately scientific accounts of these equine responses are scarce. Sympathetic horsemanship manipulates the behaviour and perceived hierarchy to encourage the horse to trust the trainer and accept him as the dominant member of the dyad.

When removed from the safety of the herd, prey animals are at far greater risk of attack from predators. Therefore as a social herbivore, the horse is safe when it is within the herd. An alpha's disciplinary action involves removing this comfort until the miscreant behaves appropriately, and asks to be let back into the herd. So by sending the naïve horse away, the trainer is putting the horse in an uncomfortable situation. The horse is not permitted to relax until he begins to focus his attention on the trainer. When this happens, the horse tends to fix an ear and eye on the trainer and may also start to chew, lick his lips and lower his head. It is unclear whether this behaviour is a sign of 'submission' or a displacement activity, but it is a reliable signal that the horse is anxious about the situation and is likely to respond to the opportunity of displaying an alternative behaviour. Hence, as soon as the horse displays this behaviour the trainer ceases to signal to the horse to move away, and instead turns his body slightly and allows the horse to approach. Repetition of this interaction between the dyad fosters the formation of hierarchical bond that precipitates a following response because the horse appreciates that it is safer and more comfortable (i.e. his comfort zone) is with the trainer. Once this lesson has been learned the horse will seek the trainer's company as a default behaviour. This response is perhaps the most significant difference between horses trained under the sympathetic rather than traditional methods. Traditionally trained horses usually remain in the vicinity of their handler largely because of being restrained by a lead rope or reins, or occasionally out of fear of being punished for leaving. Predictably, the following response makes numerous everyday horse handling tasks simple. For example, catching a horse in a large paddock becomes very easy when the horse actively seeks the company of its trainer regardless of satiation.

As well as drawing on behaviours that are familiar to horses from conspecific interactions, these methods also utilise other types of learning. As with the 'approach and retreat' method, the horse learns that its own moving away behaviour is punished both by the 'aggressive' body stance of the trainer and also the work of trotting or cantering in a small sand arena. It subsequently learns that attempts to stop 'running away' and turn towards the trainer are rewarded by the 'threat' being removed. This is, therefore, another form of negative reinforcement.

There seem to be two important causes of failure with sympathetic techniques particularly with inexperienced trainers. Confusion can occur if, after sending him away, the trainer does not recognise when the horse is ready to come in or demands too overt a response when looking to see that the horse is properly focussed. When focussing his attention on the trainer, the horse need not stand rigidly, but should be relaxed yet responsive to the trainer's body language. The inexperienced trainer may mistake a relaxed posture for a lack of attention and may feel inclined to unfairly punish the horse with persistent pressure. A lack of patience or the use of aggression by the trainer may prompt the horse to respond to him as if he was a predator. In this case the horse may become fearful (which will inhibit the appropriate operant learning), or defensively aggressive (see Figure 4).

The second potential problem is that horses may learn to offer this pattern of behaviour only in specific contexts. For example, a young horse may show exemplary behaviour in a 'round pen' but not when outside, or may do so with one trainer but not other handlers. For desirable responses to be consistent, as with all types of training, training must be generalised to a number of contextual situations.

Figure 4. The Romans used round pens (Gyrus) for training purposes.

When trained properly by this method, the horse should not only follow the trainer, but should remain focussed on him. This also facilitates future education since the horses is constantly looking for cues that help him avoid being sent away. Capitalising on the fact that horses have excellent vision, the best trainers learn to issue these cues with impressive subtlety.

The welfare of horses trained using sympathetic techniques may not necessarily be better than that of traditionally trained horses. However, the relaxed demeanour of many sympathetically trained horses and their reduced tendency to show aversive responses, including panic, means that they are easier and safer to work with. These desirable outcomes seem certain to ensure the dissemination of sympathetic techniques, such as those championed by Monty Roberts, to the wider equestrian community.

Similar principles underlie the 'natural Horsemanship' techniques proposed by Pat Parelli. As with any training programme, consistency is the key. His programmes involve a series of stages, termed 'games' through which the horse and handler progress. The early stages of these programmes develop a consistent relationship between horse and handler based on the use of body position to reward or 'punish' the horse for appropriate or unwanted behaviours. This system, again, combines stylised intraspecific communication patterns with negative reinforcement to encourage the horse to behave in the required manner.

11.6. OTHER TECHNIQUES

With the proliferation of horses as companion animals, a range of other methods of training or re-training horses have emerged over the past 20 years. Linda Tellington-Jones (1985) for example, developed a method using touch to attain co-operation with her equine patients. Gentle circular movements of the hands are used to relax the horse, using sites and strokes similar to allogrooming. The relaxed state is then associated with other stimuli, such as the voice. Certain vocal signals, as well as touch in certain parts of the horse's body, can then be used to associate new, or previously mildly aversive, stimuli, with a relaxed state. A similar kind of learning may be involved in aromatherapy treatments for horses – a particular scent is associated with a relaxed and comfortable state, and hence when that scent is reintroduced at a later stage, there is a conditioned response to a more relaxed state. Proponents of aromatherapy, however, believe that there are inherent responses by horses to scents. This may be true in some cases as it occurs in other species, such as the response to catnip in domestic cats.

12. New Equipment Used in Training Horses

12.1. SYMPATHETIC REINS

There can be nothing rewarding about pressure in the mouth. Therefore elasticised reins that compensate for clumsy riding or unyielding hands, are becoming more acceptable in equitation. It would be pleasing to see these reins manufactured with increasing levels of elasticity since some models have to undergo tremendous tension before deforming sufficiently to give horses relief. These reins are however, especially useful for jumping horses that are often jabbed in the mouth by inexperienced or incompetent riders, these reins promise to arrest habituation changes that make horses less responsive (ie hard-mouthed).

Good trainers always seek to decrease the application of aversive stimuli but this can be difficult for a third party to observe. For example when a riding teacher asks a student to reduce the tension in the rein without losing contact with the horse's mouth, there is no scientific means of ensuring that this has been followed or to what extent. The teacher has to rely on the posture and demeanor of the horse, because the reins appear constant, regardless of the tension. Especially in horse training, there is a need for a device that measures, displays and logs the tension in reins (McGreevy 1996). It has the potential to facilitate:

1. teaching of equitation;
2. consistent handling of horses by a number of riders (a programme can be followed and a senior trainer can be assured that in his absence excessive tension, which is undesirable because it can lead to habituation and unresponsiveness, will not be used);
3. research into the reactivity levels of horses. Equine behavioural scientists recognise that this element of horse behaviour can only be studied in a relevant sense by horses as they are ridden.

By identifying the key differences between a rider's use of the rein and the horse evading the discomfort, such a device could record the extent to which the horse pulls against the rider and the rider pulls against the horse. Integrated into traditional reins on either side of the horse, the innovation could offer a clear visual display which riding teachers can interpret and to which students can conform. Furthermore such devices could be used to measure the signal used by elite riders, so helping to demystify equitation skills.

12.2. TRAINING HALTERS

A number of the modern horse trainers advocate the use of a halter, training head collar or pressure halter. Similar devices have been used since horses were first domesticated. These halters work by negative reinforcement – as the horse pulls away from the handler, and the rope tightens, there is increasing pressure around the nostrils of the horse and on the poll. Hence, the horse learns that the more that it pulls, the more pain it experiences, but that when it stops pulling, the pain is relieved.

Although these head collars are effective in some circumstances, they are perhaps not as humane as some of their proponents claim them to be. In addition, there is a risk, as discussed earlier with other types of negative reinforcement, that the horse either finds an alternative way of relieving the pressure, or panics and becomes dangerous. Handlers using such devices with nervous horses need to anticipate that the horse may not always respond as desired. For example they should be aware of evasions such as lunging forward or pulling the rope away from the handler to escape. As with other types of negative reinforcement, these halters are best introduced in a non-threatening situation, where the horse can learn calmly about the effect of its own behaviour on the pressure that it feels. The device should not be used in an aversive situation, such as loading a reluctant horse into a trailer or taking it into water.

13. Conclusion

Despite the long history of horsemanship it appears that we have progressed only a little in the techniques used for training horses. Most contemporary techniques are still based upon traditional methods and negative reinforcement. The more innovative methods for training horses, such as those advocated in natural horsemanship are not new but methods that take into account the natural behaviour of the horse and use positive reinforcement are slowly developing, but are often difficult to apply in practice. As with traditional techniques, these methods would benefit from further ethological analysis.

14. References

Anderson, J.K. (1961) *Ancient Greek Horsemanship*. University of California Press, Berkeley, USA.

Berrenger, R. (1771) *The History and Art of Horsemanship*. London, UK.

Blackshaw, J.K., Kirk, D. and Creiger, S.E. (1983) A different approach to horse handling, based on the Jeffery method. *International J. Studies of Animal Problems* 4, 2.

Budiansky, S. (1997) *The Nature of Horses*. Weidenfeld and Nicolson, London, UK.

Cavendish, W. Duke of Newcastle (1743) *A General System of Horsemanship*. Newcastle, UK.

Clarke, J.V., Nicol, C.J., Jones, R. and McGreevy, P.D. (1996) Effects of observational learning on food selection in horses. *Applied Animal Behaviour Science* 50, 2, 177–184.

Clutton-Brock, J. (1987) *A Natural History of Domesticated Mammals*. Cambridge University Press, Cambridge.

Clutton-Brock, J. (1992) *Horse Power*. Harvard University Press, Harvard, USA.

Creigier, S.E. (1987) *Trailer problems and solutions*. In *Current Therapy in Equine Medicine*, volume 2. Saunders, Philadelphia, USA, pp. 135–138

Denhardt, R.M. (1947) *The Horse of the Americas*. University of Oklahoma Press, Oklahoma, USA.

Flannery, B. (1997) Relational discrimination learning in horses. *Applied Animal Behaviour Science* 54, 267–280.

Fraser, A.F. (1992) *The Behaviour of the Horse*. CAB International, Wallingford, UK.

Fraser, A.F. and Broom, D.M. (1992) *Farm Animal Behaviour and Welfare*, 3rd edition. Baillière Tindall, London, UK.

Heird, J.C., Lennon, A.M. and Bell, R.W. (1981) Effects of early experience on the learning ability of horses. *J. Animal Science* 53, 1204–1209.

Heird, J.C., Lokey, C.E. and Logan, D.C. (1986) Repeatability and comparison of two maze tests to measure learning ability in horses. *Applied Animal Behaviour Science* 16, 103–119.

Houpt, K.A. and Hintz, H.F. (1983) Some effects of maternal deprivation on maintenance behaviour, spatial relationships and responses to environmental novelty in foals. *Applied Animal Ethology* 9, 221–230.

Hyland, A. (1990) *Equus, The Horse in the Roman World.* Batsford, UK.

Hyland, A. (1994) *The Medieval War Horse.* Alan Sutton Publishing Co., UK.

Lieberman, D.A. (1993) *Learning: Behaviour and Cognition.* Brooks/Cole Publishing Company, Pacific Grove, California, USA.

Lindberg, A.C., Kellnad, C. and Nicol, C.J. (1999) Effects of observational learning on acquisition of an operant response in horses. *Applied Animal Behaviour Science* **61**, 187–201.

Littauer, M.A. (1969) Slit nostrils of equids. *Zeitschrift fur Saugetierkunde* **34**, 183–186.

Mader, D.R. and Price, G.O. (1980) Discrimination learning in horses: effects of breed, age and social dominance. *J. Animal Science* **50**, 962–965.

Mal, M. E. and McCall, C. A. (1996) The influence of handling during different ages on a halter training test in foals. *Applied Animal Behaviour Science* **50, 2**, 115–120.

Mal, M.E., McCall, C.A., Cummins, K.A. and Newland, M.C. (1994) Influence of preweaning handling methods on post-weaning learning ability and manageability of foals. *Applied Animal Behaviour Science* **40, 3/4**, 187–195.

Marshall, L.M. (1989) *Novice to Advanced Dressage.* J.A. Allen, London

McGreevy, P.D. (1996) *Why Does My Horse . . . ?* Souvenir Press, London, UK.

Miller, R.M. (1989) Imprint training the newborn foal. *Large Animal Veterinarian* **44, 4**, 21.

Miller, R.M. (1998) Imprint training the newborn foal, *Equine Veterinary Journal – Equine Clinical Behaviour Suppl.* **27**, 63–64.

Mills, D.S. (1998) Personality and individual difference in the horse, their significance, use and measurement. *Equine Veterinary Journal: Behaviour supplement* **27**.

Morgan, N. (1609) *Perfection of Horsemanship.* Edward White, UK.

Morgan (1979) *Translation of 'The Art of Horsemanship' by Xenophon.* J.A. Allen, UK.

Pearce, J.M. (1997) *Animal Learning, and Cognition: An Introduction.* Psychology Press, Hove, USA.

Press, J.M., Davis, P.D., Wiesner, S.L, Heinemann, A., Semik, P. and Addison, R.G. (1995) The national jockey injury study: an analysis of injuries to professional horse-racing jockeys. *Clinical Journal of Sport Medicine* **5, 4**, 236–240.

Prosser, C.L. (ed.) (1991) Environmental and metabolic physiology: neural and integrative animal psychology. In *Comparative Animal Psychology*, 4th edition. John Wiley, Chichester, UK.

Richardson, C. (1985) *Driving, The Development and Use of Horse Drawn Vehicles.* Batsford, UK.

Richardson, C. (1998) *The Horse Breakers.* J.A. Allen, London, UK.

Roberts, M. (1997) *The Man Who Listens to Horses.* Arrow Books, London, UK.

Roe, F.G. (1955) *The Indian and The Horse.* University of Oklahoma, UK.

Rubin, L., Oppegard, C. and Hintz, H.F. (1980) The effect of varying the temporal distribution of conditioning trials on equine learning. *J. Animal Science* **50, 6**, 1184–1187.

Seaman, S., Davidson, N. and Waran, N.K. (2002) The reliability of behaviour tests for assessing temperament in horses. *Applied Animal Behaviour Science* (in press).

Scott, J.P. and Marston, M.V. (1950) Social facilitation and allelomimetic behaviour in dogs: II The effects of unfamiliarity. *Behaviour* **2**, 135–143.

Telegrin, D.Y. (1986) *Dereivka, a Settlement and Cemetery of Copper Age Horse Keepers on the Middle Dneiper.* BAR International Series 287.

Tellington-Jones, L. (1985) *The Tellington-Jones Equine Awareness Method.* Breakthrough Publications, UK.

Tyler, S.J. (1972) The behaviour and social organisation of the New Forest ponies. *Animal Behaviour Monographs* **5**, 2.

Thorpe, W.S. (1963) *Learning and Instinct in Animals* (2nd edn.). Methuen, London

Visser, E.K., van Reenan, C.G., Hopster, H., Schilder, M.B.H., Knaap, J.H., Barnveld, A. and Blokhuis, H.J. (in press) Quantifying aspects of young horses' temperament: consistency of behavioural variables. *Equine Veterinary J.*

Voith, V.L. (1986) Principles of learning. *Veterinary Clinics of North America, Equine Practice* **2, 3**, 485–506.

Williams, M. (1997) *Horse Psychology.* J.A. Allen, London, UK, 208 pp.

Wynmalen, H. (1938) *Equitation.* Country Life, UK.

Chapter 8

WELFARE OF THE RACEHORSE DURING EXERCISE TRAINING AND RACING

D.L. EVANS
Faculty of Veterinary Science, University of Sydney, NSW 2006, Australia

Abstract. The welfare of horses in training for racing and competition can be compromised by errors of management of many processes. Lameness is usually identified, as the major problem facing horse trainers and high lameness rates in racehorses is a major welfare concern. Recent epidemiological studies have shed light on important environmental risk factors for lameness and catastrophic incidents during training and racing. Another important threat to the welfare of the athletic horse is failure of appropriate preparation of the horse for competition, resulting in earlier fatigue during a race. Fatigue during racing causes sub-optimal performance, increases the likelihood of injury and, in prolonged exercise contributes to exhaustion and even death. Failure to allow appropriate recovery periods after episodes of training and competition also contributes to a state of chronic fatigue. Trainers recognise that affected horses (or 'stale' horses) often have mood disturbances and are reluctant to exercise. Continued excessive training and inadequate recovery (termed, over-training) can result in weight loss and poor performance that is not reversed by short-term recovery periods. In events involving prolonged exercise, the performance and welfare of the horse are compromised by inappropriate fluid balance before and during exercise. Failure to properly prepare and maintain fluid balance of endurance horses results in a severe threat to welfare. Pronounced dehydration and hyperthermia can result in exhaustion and death.

1. Introduction

Racehorses are generally required to compete at maximal speeds for between one and three minutes, and endurance horses compete at slow speeds over many kilometres. Most Thoroughbred and Standardbred races occur on circular tracks, with horses required to travel at 50–60 kilometres per hour. Competition at these speeds results in very high metabolic rates, and high impact forces on the limbs. It is not surprising that injuries are common in racehorses, and lameness is the most common health problem and cause of wastage. In endurance horses, the challenge to health is mostly related to the demands of maintaining body temperature and body fluid status.

In this chapter results of recent studies of risk factors for lameness during racing is reviewed. As well, racing of two year old Thoroughbreds is discussed in the light of recent studies, and the overtraining syndrome discussed. Recent studies on mechanisms of exercise-induced pulmonary haemorrhage and appropriate training of horses for competition, are reviewed. Strategies for appropriate fluid and electrolyte balance in endurance horses are also considered.

N. Waran (ed.), The Welfare of Horses, 181–201.
© 2002 *Kluwer Academic Publishers. Printed in the Netherlands.*

2. Musculoskeletal Injuries and Lameness

Musculoskeletal injuries are very common in athletic horses, and lameness is probably the greatest cause of concern for the welfare of horses in training for racing and eventing (see Figure 1). There have been several studies of the epidemiology of lameness in racehorses, and this section illustrates the scope of the problem. It is important that studies of this type are conducted because solutions to the problem of high lameness rates are unlikely to be found if the nature of the problem is not clearly described and understood by all persons and organisations with an interest in the welfare of the racehorse.

MInjuries and disease sustained in training are less visible to the public than racing injuries, yet they represent an important source of wastage to the industry (Jeffcott *et al.*, 1982) and raise similar concerns for animal welfare. Whilst racing injuries are officially recorded by veterinarians employed by race-clubs in Australia, official records of injuries in training are not maintained. Trainers are not usually required to report injuries sustained during training and the veterinarians servicing the stables are not compelled to notify the stewards of cases they have treated. In contrast, the regulation of veterinary services by the racing authorities in Japan and Hong Kong makes it possible to maintain central records of injuries sustained in training (Watkins 1985; JRA 1991).

Maintenance of detailed records of injuries sustained during training and after racing is fundamental to the process of investigating the causes of the problems,

Figure 1. Racing imposes certain risks of injury on the horse.

and measuring the success or failure of changes to factors such as training, racing, and track design and reconstruction. It is unfortunate that large sums of money are frequently spent on redesign and construction of racetracks in the absence of either good evidence for the changes, or studies of the effect of the changes. Racetracks for Standardbred racing have undergone radical changes in design in many countries. These changes to racetrack construction are an important contributor to improved horse welfare because increased banking of the turns significantly reduces lameness rates after races on these tracks (Evans & Walsh 1997).

Most Thoroughbred horses in Australia are trained on racetracks and must therefore gallop around corners during training. Research has shown that the strain on the third metacarpal bone increases when turning a corner, and increasing the tightness of the corner increases the strain. The strain on the outside limb was consistently more than on the inside limb when exercising around a turn. This difference is accentuated when speed is increased (Davies 1996).

Standardbreds race at approximately 800 metres per minute, often on small racetracks with a circumference of 800 metres. The strains on the limbs will be increased if the design of the racetrack includes irregularities in the surface, and if the turns are not properly banked, as in a cycling velodrome. Underbanking on the turns of a racetrack refers to a design that results in excessive centrifugal forces on the horse as it travels around the bend. Investigations of racetrack design for Standardbred trotters have confirmed the need for adequate banking. When a racetrack was redesigned to remove underbanking in semicircular curves, improve the transition between curves and remove sloping sections on the straight there was marked reduction in gait asymmetry and the heat in fetlock joints (Fredricson et al., 1975a, b). These results suggest the strain on the limbs had been reduced when the banks were steeper on the corners (Fredricson et al., 1975a, b).

The results of these studies indicate that track design and banking of corners may also be a major factor contributing to injuries in Thoroughbred racehorses. Increasing the radius of corners and the degree of banking and placement of inclines in straight stretches may be useful in reducing low-grade lameness such as shin soreness. In addition, during the early stages of training, it would also be advisable to reduce the speed at which horses travel around corners.

Increasing the load a horse must carry increases the ground reaction forces. This increase in ground reaction forces could increase the load placed on the bone during training. Therefore it is possible that the use of heavier riders could increase the incidence of injuries. The skill of the rider may also affect the forces placed on the third metacarpal bone. Studies in trotting horses have demonstrated that a rider is able to redistribute weight borne by the horse from the front limbs to the hindlimbs (Schamhardt et al., 1991). Training without a rider, for example, on a treadmill, may help reduce lameness rates. However, treadmill training is not specific for the demands of racing, and should not be used exclusively.

Reduced lameness rates could be a factor in the superior earnings from prize money in Thoroughbred stables in USA that use treadmills as part of their routine training. A large-scale study has studied racing performance of Thoroughbreds in North America with a history of treadmill training (Kobluk et al., 1996). Horses that had been trained on the treadmill for at least 50% of their program for at least 60 days prior to the start of racing were defined as treadmill trained. Racing performance in 107 treadmill-trained horses was compared with results in 214 control horses trained conventionally. In all age groups and classes the treadmill-trained horses were equal or superior to the conventionally trained horses.

A study examining all types of lameness in a number of different racing stables found that lameness decreased in stables with higher numbers of horses and increased in stables with higher levels of veterinary and farrier involvement (Ross & Kaneene 1996). It is not clear what effect these factors would have on the incidence of shin soreness. Furthermore these results could reflect differences in rates of injury detection, rather than rates of lameness.

Bone mineral density contributes to bone strength, and it can be measured accurately in the distal limb of growing or training horses. Training results in increased mineral density (Firth *et al.*, 2000). A study comparing pasture raised foals to those raised in boxes with and without additional exercise found that pasture raised foals appeared to have a "stronger" musculoskeletal system (Weeren *et al.*, 2000). However, most of the difference in bone mineral density between the three groups disappeared after six months without exercise. Furthermore, as this study did not follow the horses beyond 11 months it is not possible to draw conclusions regarding the permanent damage and consequences for later performance of housing during early development. Differences in exercise levels in foals also had implications for development of articular cartilage, tendon and muscles, and can influence gait. Enforced withholding of exercise retarded development. The authors reported that absence of exercise in the foals had a long-term effect on collagen characteristics in articular cartilage. They speculate that it is possible that this effect could have implications for injury resistance later in life. The authors recommend pasture raising for foals in the first year of life.

2.1. INJURIES IN TWO YEAR OLD THOROUGHBREDS

The training and racing of two-year-olds is a contentious issue because of welfare concerns about subjecting relatively immature horses to high work demands. Mason and Bourke (1973) followed 74 Thoroughbreds in Australia during their two-year-old racing season and reported that 40% were unsound at the end of the season. The most common cause of unsoundness was sore shins (46% of cases). Shin soreness is a training injury that affects the upper dorsal aspect of the third metacarpal bone. The complaint is most commonly observed in young Thoroughbred and Quarter horses during their first racing preparation. The condition is characterised by pain on palpation of the metacarpus and is often associated with an unwillingness to work at high speed. Continued training may lead to diffuse soft tissue swelling visible on the dorsum of the metacarpus (Stover *et al.*, 1988).

Approximately 9% of cases of lameness were associated with sore shins in the UK (Jeffcott *et al.*, 1982; Rossdale *et al.*, 1985). A survey of veterinarians and trainers estimated that shin soreness affected 80% of two-year-olds in Australia (Buckingham & Jeffcott 1990) and 70% in the United States (Norwood 1978). The higher frequency of shin soreness in Australia compared to the UK may be due to the greater emphasis on two-year-old racing in Australia. It also may be associated with the training of horses on tracks involving turns in Australia, unlike many of the straight training tracks in the UK, because strains increase on the cannon bone when horses are exercised around a turn (Davies 1996).

Knee problems (17%), splints (13.5%), sprained fetlock joint (9.5%) and sesamoid problems (5.4%) were also common in the study by Mason and Bourke (1973). The fetlock joint of the front leg is susceptible to injury because it has a relatively small surface area, it has the greatest range of motion of any of the limb joints, and flat racing horses transmit all of their weight through this one joint during one phase of the stride (Pool & Meagher 1990).

In a study of the racing careers of 353 horses sold as yearlings in Australia (Bourke 1995), it was reported that horses that raced as two-year-olds had a greater number of starts over a lifetime and raced in more seasons than those racing first at three years of age. This suggests that there is no detrimental effect of starting racing at two-years of age. However, some horses that first raced in later years may have been given time to mature. Others could possibly have entered training as two-year-olds but had their racing debut delayed because of injury or a lack of ability (Physick-Sheard 1986), and both these factors could subsequently limit the horses' careers. Racing could also be delayed if horses had poor conformation. In the United States, a higher proportion of two-year-olds sustain an injury compared to older horses (Robinson & Gordon 1988), and in Germany, a larger proportion of two-year-olds have training failures compared to older age groups (Lindner & Dingerkus 1993). A similar situation exists in the UK, where two-year-olds lose a greater proportion of available training days due to lameness compared to three-year-olds (Rossdale et al., 1985).

One of the major aims of an epidemiological study by Bailey (1998) was to document the time lost in training due to various categories of injuries and disease. This information could then be used to determine the relative importance of lameness and respiratory conditions, and to determine to what extent injuries and disease in training are responsible for the high proportion of elite horses that do not race as two-year-olds in Australia.

From 525 yearlings catalogued at a major yearling sale in 1995, 169 horses placed with 24 participating trainers were enrolled in a long-term study designed to identify causes of wastage to the Thoroughbred industry (Bailey 1998). Horses were followed from the time of sale until the end of the cohort's three-year-old racing season. Records were maintained on the training, injury and disease status of the horses in the cohort.

Of the 169 horses included in the study, 160 (95%) had entered training in the stable by the end of the two-year-old season, whereas only 76 (45%) had raced as two-year-olds. Eighty-five per cent of the horses suffered at least one incident of injury or disease whilst in training as a two-year-old. The most common injury in two-year-olds was shin soreness, which affected 42% of the 160 horses, followed by fetlock problems (25%) and coughs and nasal discharges (16%). Thirteen per cent suffered from cuts or traumatic injuries, 9% from foot problems, 7% from knee problems, 6% from tying up, 5% from ligament sprain, 3% from fever of unknown origin, 2% from upper respiratory obstruction (*e.g.* roarers) and 1% from tendon strain. Of the horses that suffered from shin soreness as two-year-olds, 40% developed shin soreness a second or third time as a two- or three-year old. The corresponding figure for recurrences of fetlock problems was 48%, and for coughs and nasal discharges 27%.

Bailey (1998) reported that lameness, excluding cuts and traumatic injuries, was the most important veterinary cause of lost training days during the study period (56.2% of total days modified), followed by respiratory conditions (15.8%). These results support those from studies on Thoroughbreds in the UK (Jeffcott *et al.*, 1982, Rossdale *et al.*, 1985) and Germany (Lindner & Dingerkus 1993). This finding was not surprising given that lameness encompasses a wide range of problems and the musculoskeletal system is subjected to frequent stresses from training and racing (Pool & Meagher 1990).

This study of horses in training by Bailey (1998) has provided objective information on the relative impact of injury and disease in Australian Thoroughbreds. Although there is considerable emphasis on two-year-old racing in Australia, less than 50% of elite horses raced during this year. The principal reason for this low figure was the high number of cases of low-grade injuries and disease that occurred during the training of two-year-olds. These minor incidents often altered training or resulted in the horse being rested at pasture, but did not prevent the horse from racing in subsequent seasons. In contrast, major injury was relatively uncommon in young horses in training. Therefore, whilst major injury is relatively uncommon in young horses in training, low-grade injury and disease have the potential to disrupt training schedules, cause significant economic loss and are an important welfare concern.

These findings emphasise the need for further studies of the risk factors for lameness. Epidemiological studies that investigate the multifactorial nature of lameness have contributed important knowledge already, but more studies are needed. It is likely that results of studies in one training environment will not be directly applicable in another because training facilities and practices differ widely. Studies of trotters, pacers, endurance, event and Quarter horses are also needed.

2.2. SHIN SORENESS

Shin soreness is a major cause of lameness in young Thoroughbred horses. The inflammation of the dorsal aspect of the metacarpus is a major animal welfare concern, and a cause of industry wastage. In the following section this disease will be discussed in more detail. The normal responses of bone to the forces that occur during exercise are outlined, and the current knowledge of risk factors for the disease is discussed.

When horses enter training the thickness of the third metacarpal cortical bone increases (Davies *et al.*, 1999). This thickening helps the bone to resist bending forces in the dorsal palmar direction (Nunamaker 1996). Bone remodelling involves changing the internal architecture of the bone cortex without altering the shape of the bone. The remodelling process allows the bone to adapt to mechanical loading and is thought to prevent the accumulation of micro fractures. During the remodelling process a small packet of bone tissue is reabsorbed and new bone is deposited in its place. At the completion of re-modelling the bone requires approximately three months to mineralise sufficiently (Pool 1991; Nunamaker 1996; Brunker *et al.*, 1999).

Following reabsorption there is a period of 1–2 weeks before new bone formation commences. During this period the bone density at that site is reduced and the bone is weakened. Continued loading during this period may result in micro damage accumulation and the beginning of overuse injuries, such as shin soreness (Riggs & Evans 1990; Nunamaker 1996). It has been demonstrated in Quarterhorses that bone density decreases significantly during the first 2 months of training and does not begin to increase until after nearly 3 months of training (Nielsen *et al.*, 1997). The duration of training in many establishments is therefore unlikely to be sufficient for optimal adaptation of bone prior to racing.

The application of load, below that required to cause complete fracture, produces micro damage. If this load is applied repeatedly the level of micro damage will increase and the remodelling process will be accelerated. At a microscopic level the first signs of accelerated remodelling are vascular congestion, thrombosis and resorption of bone. In humans accelerated remodelling does not, initially, produce any symptoms. However as remodelling progresses mild pain will occur during exercise. If loading does not stop then the pain will persist even after the completion of exercise (Brunker *et al.*, 1999).

If the bone loading continues then the size of the resorptive cavities increases, resulting in the appearance of micro fractures that extend into the cortex. These cracks cause a marked reduction in bone strength. At some point there may be insufficient bone to withstand the load. If this happens then a complete fracture occurs (Riggs & Evans 1990; Nunamaker 1996).

The incidence of shin soreness is highest in two-year-old racehorses (Moyer *et al.*, 1991, Bailey 1998). However, there has been no investigation to determine if the age at which training commences influences the incidence of shin soreness. Bailey (1998) found that overall injury rates were no different between males and females. There has been no work conducted to determine if conformation plays a role in the development of shin soreness.

Bone strength at the commencement of training may be an important factor affecting the likelihood of developing shin soreness. The strength of the bone will influence its ability to withstand repeated loading. The total strength of the bone is determined by its stiffness or elasticity, mineral density and shape. Bone density can be estimated by examination of x-rays. However this technique is not precise. Precision can be improved with photon absorptiometry. Combination of photon absorptiometry and ultrasound velocity can be used to estimate the bone elasticity (Jeffcott *et al.*, 1988). It is possible to determine the shape of the bone by manually measuring the cortical thickness from x-ray, although the technique is highly subjective (Brunker *et al.*, 1999).

Bone geometry, or cross sectional area is an important factor in determining the

Table 1. Distances exercised per week in five stables for horses that did and did not develop skin soreness (adapted from Boston & Nunamaker 2000).

| | | Stable | | | | | |
		1	2	3	4	5	Average
Total distance	Shin sore	15568	6880	8160	16048	11712	11904
(meters/week)	Not shin sore	12144	8624	9424	14512	14928	10704
Jogged	Shin sore	5296	1232	1904	5856	8288	3744
(meters/week)	Not shin sore	4512	1712	2272	5344	9968	3568
Galloped	Shin sore	10064	5296	6112	9952	3376	7952
(meters/week)	Not shin sore	7424	6384	7024	8928	4768	6880
Breezed	Shin sore	208	352	144	240	48	208
(meters/week)	Not shin sore	208	528	128	240	192	256

load a bone can withstand before failing. In Quarter horses greater cortical mass in the dorsal and medial metacarpal at the commencement of a training program has been associated with a lower injury rate (Nielsen *et al.*, 1997).

An inability to attain a previous maximum exercise speed in a subsequent run, within the same exercise period, is associated with an increase in bone strain (Davies 1996). Therefore the ability of a muscle to withstand the demands of exercise and resist fatigue could be an important factor in the development of shin soreness. These results emphasis the importance of training techniques that maximise the adaptations of exercising muscle that facilitate coping with the demands of competition.

Training methods are an important risk factor for shin soreness in Thoroughbreds. A study in the USA found that varying the speed and distances galloped during "fast work" could reduce the incidence of shin soreness. In this study training speeds were divided into breezing (approximately 900–960 m/min), galloping (approximately 660 m/min) and jogging (approximately 300 m/min). The average distance breezed and galloped in each stable is shown in Table 1. The results showed that the risk of developing shin soreness increased as the weekly distance galloped increased. However the risk decreased as the weekly distance breezed increased. Furthermore the weekly distance jogged did not affect the risk of horses becoming shin sore. The authors concluded that in order to reduce the incidence of shin soreness, trainers should allocate more training time to regular short-distance breezing and less to long distance galloping. However the distance breezed and/or raced should not exceed 5000 m in a 2 month period because it could increase the risk of a fatal injury (Boston & Nunamaker 2000).

Further research needs to be conducted to determine training-related risk factors for shin soreness in other countries, where training techniques are likely to be different.

Training surfaces have long been considered a contributing factor to shin soreness (Buckingham & Jeffcott 1990). Hard tracks are a problem because the ground reaction forces are increased, thereby increasing the strain on the bone. Alternatively training on soft surfaces, whilst providing cushioning, may hasten muscle fatigue (Brunker et al., 1999), which would also increase the strain on the bone (Davies 1996).

Training on wood fibre surfaces may decrease the occurrence of shin soreness (Moyer *et al.*, 1991; Moyer & Fisher 1992). Thirty four percent of horses trained exclusively on dirt developed shin soreness compared to 13.5% of horses trained on woodchip. In addition the horses trained on woodchip accumulated 86 miles of fast work before the onset of shin soreness, whilst horses trained on dirt accumulated only 32 miles. It should be noted that the horses were trained not only on different surfaces but also at different training centres. It is therefore possible that the different injury rates were due to factors other than the differences in the surfaces, such as track geometry.

Dietary deficiencies, especially in calcium, may influence bone density and remodelling. Young Quarter horses entering training and fed high levels of calcium (34.9 grams/day) and phosphorous (26.4 grams/day) had a higher bone density in the third metacarpal bone than those feed 28.3 grams/day of calcium and 21.9 g/day of phosphorous (Nielsen *et al.*, 1998). It is not known if increasing the level of calcium and phosphorous will reduce the risk of shin soreness in Thoroughbred or Quarter horses.

2.3. RISK FACTORS FOR CATASTROPHIC INJURY IN RACEHORSES

Estberg *et al.* (1998) investigated the relationship between intensive racing and training schedules, and risk of either catastrophic musculoskeletal injury (CMI) or lay-up from racing in Californian Thoroughbreds. Periods of rapid average daily accumulation of high-speed exercise distance were identified for each horse from official race and training histories. Horses that had a period of rapid accumulation of high-speed exercise distance (a hazard period) had 4.2 times the risk of CMI within 30 days. Horses were also 4.8 times more likely to have to be spelled (laid up) after a period of rapid accumulation of high-speed exercise distance. In summary, rapid increases in distance of high-speed exercise in Thoroughbreds increases the likelihood of a catastrophic musculoskeletal injury (CMI) and having to 'spell' the horse.

The association between distances of high-speed exercise and injury has also been investigated in Thoroughbreds in Kentucky (Cohen *et al.*, 2000). This study found no association between high-speed exercise and onset of either catastrophic or non-catastrophic injury. In fact, cumulative high-speed exercise was inversely associated with risk of injury. This means that the more exercise a horse accumulated, the less likely it was to become injured. For example, a horse with 10 more cumulative furlongs of high-speed exercise during the 1 month period before a race than another horse was nearly 2 times less likely to be injured. These results also suggest that in order to adapt a horse for safe racing, it is important to give high-speed exercise during training. Lack of high-speed exercise in training may cause decreased skeletal density, and consequent increased risk of injury. These results conflict with those reported for Californian Thoroughbreds (Estberg *et al.*, 1998). Cohen *et al.* (2000) suggested that there might be regional differences in the role of high-speed exercise as a risk factor for injury. In addition, the association of increased risk of injury with accumulated high speed exercise in the Californian study might reflect pre-existing health problems or injury that limited the ability to perform high speed exercise. Trainers might try to limit high-speed exercise in horses that have existing injuries, or have had previous lameness.

These findings by (Cohen *et al.*, 2000) illustrate the importance of specificity of training. Horses are more likely to cope with the demands of racing if the training includes appropriate use of the workouts at the speeds used in races. Of course these high-speed workouts should be introduced gradually, and used at appropriate frequencies.

Horses should be cautiously reintroduced to fast exercise after a 'spell' from training. The importance of care in the three-week period after a 'spell' has been confirmed in a study of Californian Thoroughbreds (Carrier *et al.*, 1998). The study investigated whether a two-month or longer period without official high-speed workouts was associated with humeral or pelvic fracture within hazard periods of 10 and 21 days following a lay-up. The study investigated many aspects of race training of horses that had been euthanased because of a complete humeral or pelvic fracture. Risk factors investigated included age, sex, activity, number of lay-ups, number of days from a race or official timed workout to fracture, number of days from end of last lay-up to fracture, mean duration of lay-ups, and total number of days in race training. Horses with pelvic fractures were more often female, older, and had either no lay ups or greater than or equal to 2 lay-ups. Horses with humeral fractures were typically 3-year-old males that had had one lay-up. Horses with pelvic fractures had more total days in race training, fewer days from last exercise event to fracture, and a greater number of days from end of last lay-up to fracture, than horses with humeral fractures. Return from lay-up was strongly associated with risk for humeral fracture during hazard periods of 10 and 21 days. It was concluded that risk of humeral fracture might be reduced if horses are cautiously reintroduced into race training after a lay-up.

Risk factors for catastrophic musculoskeletal (MS) injury in Thoroughbred racehorses racing on dirt and grass racetracks in Florida has also been reported (Hernandez et al., 2001). The incidence of MS injury overall was 1.2/1000 race starts. Higher risk of injury was associated with more than 33 days since the last race, and turf-racing surface. Geldings also had higher risk. The association with the number of days since last race may reflect a previous injury or health problem. Higher risk on grass tracks in this study could have been due to turf races, with larger fields and higher prize money being more competitive than races on dirt. The higher risk for geldings may be associated with their racing more frequently or having more races per career than fillies and mares.

Eighteen fractures were reported in a group of 209 thoroughbred racehorses studied over a 12 month period in the UK (Pickersgill & Reid 2002a). The racing fracture incidence rate was over 16 times greater than the incidence rate on training days. No relationship was identified between either age or gender and the occurrence of fractures. Fractures are a significant cause of wastage in the Thoroughbred flat racing population. The authors also concluded that the results were suggestive of a protective effect of training horses on equitrack surfaces. The increased use of such a surface might reduce the incidence of both racing and training injuries. Further investigations of the relationships between training track usage and injury rates during training and racing are urgently needed. Such studies have the potential to change the strategies for design and use of training and racing tracks. Studies of extrinsic risk factors for injury and adoption of sensible findings by industry participants should promote racehorse welfare.

2.4. TENDON INJURY IN NATIONAL HUNT THOROUGHBRED RACEHORSES IN THE UNITED KINGDOM

The prevalence of tendinitis in one National Hunt racehorse stable in the United Kingdom was very high. Twenty five of 96 horses in the study developed tendinitis of the superficial digital flexor tendon over a 12 month period (Pickersgill *et al.*, 2002b, in press). Only 55 (57%) showed no evidence of tendinitis. Other cases included 12 horses that had chronic tendinitis, and 4 that had acute exacerbations of chronic injuries. The likelihood of a horse developing tendinitis increased with age in males and females. However, females were more likely to develop tendinitis at all ages.

2.5. EXERCISE-INDUCED PULMONARY HAEMORRHAGE

Exercise-induced pulmonary haemorrhage (EIPH) occurs in horses that compete at high speeds, including Thoroughbreds, Standardbreds and Quarter Horses. It is a major welfare issue, because horses can bleed to death during a race, or fall during a race and present a danger to other horses and their riders or drivers. The sight of a racehorse bleeding profusely from the nostrils during or after a race frequently arouses great concern about the welfare of the horse and racehorses in general. EIPH is also a likely cause of inflammation in the lower airways (McKane & Slocombe 1999) and so may contribute to poor performance and coughing in racehorses.

It is generally accepted that the cause of EIPH is rupture of the thin membrane that separates the pulmonary blood from the pulmonary alveoli and airways. Rupture is attributed to the high blood pressure that occurs normally in the pulmonary capillaries during intense exercise. Stress failure of the pulmonary capillaries occurs at pulmonary capillary pressures of 75 to 100 mmHg. These high pressures are to be expected in the pulmonary blood vessels in strenuously exercising horses (West & Mathieucostello 1994). Lung trauma in galloping horses resulting in shock waves in the thorax has also been suggested as a cause of EIPH (Schroter *et al.,* 1998).

At heart rates near maximal (in the range from 192–207 beats/min), mean pulmonary arterial pressures (mPAP) ranged from 80–102 mmHg during treadmill exercise (Meyer *et al.,* 1998). A positive relationship occurred between the number of red blood cells in the bronchoalveolar lavage (BAL) fluid and mPAP. In addition, the amount of haemorrhage increased as the mPAP exceeded 80 to 90 mmHg. These results suggest that all strenuously exercised horses may exhibit EIPH. The amount of haemorrhage appears to be associated with the magnitude of the high pulmonary arterial pressure (Meyer *et al.,* 1998). Given that up to 90% of racehorses exhibit EIPH following sprint exercise, the condition could be regarded as a normal response to fast exercise, rather than as a disease.

3. Welfare of Event Horses Competing in Hot, Humid Climates

The Atlanta Olympics in 1996 posed a serious threat to the welfare of the horses competing in the 3-day event. Anticipated conditions were 31–36 °C with relative humidity approaching 100%. Research in the years before the event provided a scientific basis for modifications to the event. Jeffcott & Kohn (1999) have reviewed this research effort, and its contributions to the welfare of competing horses. The importance of proper acclimatisation, avoidance of dehydration and use of appropriate aggressive cooling strategies are emphasised. In Atlanta the duration of all phases of the event were reduced so that the thermal load was decreased. These modifications coupled with appropriate housing, monitoring, and use of misting fans, ensured that no horses experienced heat related problems during the competition.

Methods of cooling of heat stressed horses have sometimes been a contentious issue. Some horse owners believe that application of cold water to a horse's body may cause sore muscles. There is no basis for this belief. Aggressive cooling of the whole body with cold water at 9 °C promotes the welfare of heat stressed horses (Williamson *et al.*, 1995). Rectal temperatures decreased more quickly than in horses treated with tepid water at 31 °C, and no signs of muscle disease were noted in the aggressively cooled horses.

Given that many of the horses coming to Atlanta were to be flown in from areas with a temperate climate, several groups undertook studies investigating the effects of acclimation to heat and humidity. Geor *et al.* (1996) found that as little as five days exposure to heat and humidity for 4 hours/day resulted in a lower core temperature and reduced fluid losses in response to a standardised exercise test as compared with those not exposed. These findings were supported by the studies conducted at the Animal Health Trust (Marlin *et al.*, 1996), where it was shown that horses underwent considerable physiological adaptation in terms of heat tolerance in response to daily exposure for 21 days to conditions of high ambient temperature and humidity. Horses transported for competition in hot environments should therefore be given an appropriate period of acclimatisation.

4. Endurance Horses

Failure to adequately lose body heat can be a serious threat to the health and welfare of endurance horses competing in hot, humid conditions. Fluid losses can be 12–15 litres per hour when it is very hot. Horses can lose 40 kg body weight during an endurance ride. Most of the loss is as sweat, which is rich in sodium, potassium and chloride ions.

Unless the losses of water and electrolytes are replaced, horses will become dehydrated and develop electrolyte imbalances. These responses decrease performance, and depress the horse's thirst. Dehydrated horses are also less able to efficiently regulate their body temperature and are more likely to develop severe, life threatening hyperthermia (Geor & McCutcheon 1998). Electrolyte losses contribute to development of 'thumps' (synchronous diaphragmatic flutter), muscle cramping, and 'tying-up' (exertional myopathy). Severe dehydration and electrolyte losses cause exhausted horse syndrome.

There are sensible strategies that help to avoid these threats to the welfare of endurance horses competing in hot humid conditions. Horses in regular training should receive salt (sodium chloride) in the feed. Depending on environmental conditions and sweat losses during training, 50–125 grams per day has been recommended, with free access to water. Salt blocks cannot be relied on to provide every horse with the correct intake of sodium chloride (Jansson & Dahlborn 1999). Access to good quality hay should provide sufficient potassium.

Before and during a ride an electrolyte paste administered with a large syringe can be used to stimulate drinking (Sosa Leon *et al.*, 1998). In a treadmill study, Arabian horses were given either water or an electrolyte paste before and during 60 km (38 miles) of treadmill exercise (D'sterdieck *et al.*, 1999). The exercise was divided into four bouts of 15 km. Ninety minutes before exercise, the supplemented horses were given 75 grams of a 2:1 'salt-Lite' salt/water mixture, with further 38–50 gram doses given at rest breaks between exercise bouts. Horses given the electrolyte supplement consumed twice as much water, lost less weight during exercise and maintained higher blood sodium and chloride concentrations. Horses should be offered water at every opportunity during endurance rides, including immediately after exercise. However, horses cannot be relied on to rehydrate fully by drinking in the 24 hour period after an endurance ride (Schott *et al.*, 1997).

5. Suboptimal Training as a Threat to Horse Welfare

The welfare of athletic horses is promoted by strategies that maximise fitness and reduce the likelihood of fatigue and injury. Excessive training of horses and racing inadequately trained horses are both major threats to horse welfare. There have been many studies of the physiology of exercise and training in horses. These studies have provided important information that can help promote fitness for competition. The following section summarises some research findings that have provided practical information that can be used to refine horse training in order to maximise fitness.

Training of Thoroughbred and Standardbred horses depends on using strategies that promote stamina, acceleration, and speed. Stamina is promoted by use of an initial phase of training at slow to moderate speeds that builds endurance and promotes increases in maximal oxygen uptake. During this initial phase of training occasional use of higher speed work over very short distances (200–300 m) promotes adaptations in bone. These adaptations will help prevent shin soreness later in the preparation.

Preparation of racehorses and event horses for racing or competition necessitates gradual increases in the speed of exercise. Horses racing over 400–4000 metres and the cross country phase of the second day of a three day event all experience an accumulation of lactic acid in muscle cells, and in the blood. This implies that some anaerobic metabolism is involved in the ATP resynthesis during the competition or race. It is likely that anaerobic metabolism in a Quarter horse race supplies most of the ATP. In events of 1000–4000 meters distance, anaerobic metabolism probably only supplies 20–30% of the energy (Eaton *et al.*, 1995). Endurance rides do not stimulate anaerobic glycolysis, and training of endurance horses should involve continuation of training at slow speeds similar to those used in competition.

5.1. MONITORING EXERCISE INTENSITY DURING TRAINING

There have been several studies that illustrate that superior adaptations to training can be attained if careful monitoring of exercise intensity is used. Studies of the physiological responses to training have used two methods of monitoring the exercise intensity that can be applied by trainers. These methods describe the intensity by measurement of heart rate during the exercise, or by measuring blood lactate concentration during or immediately after the exercise. Exercise speed resulting in a heart rate of 200 beats/minute has been suggested as suitable for race training (Gysin et al., 1987). However, heart rates in horses during Standardbred and Thoroughbred races are maximal, between 210 and 240 beats per minute, and excessive training at a heart rate of 200 beats per minute might not adequately prepare a racehorse for the physiological and metabolic demands of racing.

5.2. TRAINING TO MAXIMIZE THE ADAPTIVE RESPONSES

Intensity and duration of exercise training influence the rate and degree of adaptation to treadmill exercise training. Training strategies, such as interval training at intensities near maximal, can result in beneficial adaptations in muscle. Such training necessitates careful monitoring of the intensity of exercise. Exercise at sub-optimal intensities will limit the rate of adaptation, and frequent training at intensities above optimal will risk onset of fatigue and overtraining syndrome.

Training methods influence the adaptation of muscle to training. Interval training at high speeds on a treadmill resulted in increased concentration of lactate dehydrogenase (LDH) in skeletal muscle, but conventional training does not have the same effect (Lovell & Rose 1991). Lactate dehydrogenase concentration in skeletal muscle has been used as a marker of anaerobic enzyme activity. In that study Thoroughbred horses were trained for 12 weeks. In the final 3 weeks, horses exercised at a velocity that resulted in 100% of maximal heart rate for 600 metres on 3 days per week. Speeds of exercise were 9–12 m/s on a treadmill inclined at 10%. Three bouts of exercise were given, separated by recovery periods that were three times the duration of the 600 metres of exercise. On three other days, horses exercised over 3000–4000 metres at 6–7 m/s (a canter). Total distances exercised in the three weeks of training were 18, 20 and 16 kilometres. This study shows that interval training on a treadmill at speeds calculated with the help of a heart rate meter did promote superior fitness.

Training at a relative intensity of 80% of maximal oxygen uptake (VO_{2max}) for 6 weeks did not result in increases in skeletal muscle (gluteus medius) LDH concentration (Sinha et al., 1991). However, the training did significantly increase the muscle buffering capacity by 8% and increase the ratio of fast twitch, highly oxidative fibres to fast twitch fibres (FTH/FT) (Sinha et al., 1991). These adaptations to training did not occur in a group of horses trained concurrently at a lower intensity (40% VO_{2max}). These results show that superior fitness will be attained with training at higher intensities, approaching those that result in maximal oxygen uptake. At these velocities, horses will be exercising at approximately 90% of maximal heart rate. They will also have slightly elevated blood lactate concentrations, in the range 3–8 mmol/l. These results are relevant to horses being trained for Thoroughbred and Standardbred races, and probably to training of event horses.

Design of suitable training programs depends on an understanding of the metabolic demands of the event. Munoz *et al.* (1999) investigated the cardiovascular and metabolic responses to two cross-country events (CC* preliminary level and CC*** advanced level) in 8 male eventing horses. Plasma lactate response exceeded the 'anaerobic threshold' of 4 mmol/L, reaching a maximum level of 13.3 mmol/l. Heart rates ranged from 140 to more than 200 bpm, peaking at 230 bpm. The authors concluded that muscle energy resynthesis during a cross country event is provided by oxidative metabolism and glycolysis. The training of event horses should be monitored to ensure that some of the exercise results in heart rates and post-exercise blood lactate concentrations that are likely to be experienced during competition.

How often should horses exercise in order to obtain a training effect? There have been very few studies of the importance of frequency of training. However, Gottliebvedi *et al.* (1995) found that interval training at VLa_4 on only three days per week is sufficient to cause adaptational changes in exercise tolerance related parameters. VLa_4 is the velocity approximating that, which results in a blood lactate concentration of 4 mmol/l. The results also indicated that some adaptations due to training are rapidly lost over a four week period when horses cease training.

Sprint training at or near racing speeds over 400–1600 m or more represents the final phase of training for racing. The sprint training should not be combined with long duration and distances of basic training. If a high volume of base training is combined with sprint training there is a much greater risk of overtraining, resulting in poor food intake, loss of weight, injuries and disinterest in training and racing.

It is not necessary to monitor heart rate or blood lactate after fast exercise or sprint training in order to specify exact training speeds. All horses will have maximal heart rates (210–230 beats per minute), and have high blood lactate concentrations (15–25 mmol/L) after such exercise. The horse should be allowed to learn how to gallop, pace or trot at high speeds, and then the distance gradually increased. High-speed sprints need not exceed 800 metres distance. After 800 metres exercise at maximal velocity the blood lactate concentrations are similar to those measured after racing over longer distances, implying that exercise at peak speeds is probably unnecessary over distances greater than 800 metres.

5.3. USE OF SWIMMING EXERCISE

Studies of heart rates in swimming horses indicate that free swimming is similar in intensity to trotting and slow cantering (Murakami *et al.*, 1976). A training effect was found, as heart rate during swimming decreased over a four week period of regular swimming exercise. Heart rates ranged from 140–180 beats per minute, and blood lactate concentrations only increased by 2–4 fold above resting values during swimming. Horses were exercised for 5 minutes daily in the first week, and the duration was increased by 5 minutes each week thereafter. It was concluded that swimming was appropriate for the development of basic physical fitness and for rehabilitation of horses with limb problems.

Prolonged swimming for one hour did not cause excessive increases in body temperature. It was suggested that the direction of swimming in circular pools be changed regularly during prolonged swimming to avoid fatigue in the outside legs.

Some horses seem to have great difficulty breathing during swimming, and sometimes horses seem to be weak after leaving the pool. Blood pressures in the pulmonary blood vessels are also high during swimming. It is not known if these blood pressures are sufficiently high to cause frequent exercise-induced pulmonary haemorrhage (EIPH). However, trainers have reported seeing epistaxis (nosebleed) occasionally after swimming of horses.

Swimming training may reduce the frequency of lameness, even though the exercise is not specific to normal equine competitions. A Japanese study investigated whether or not swimming training changes the frequency of locomotor diseases in two year old Thoroughbreds (Misumi et al., 1994). In this study, 24 horses were divided into three groups: Group A, trained by only running; Group B, trained by running plus a gradual increase in swimming, and Group C, trained by running plus constant swimming. Only in Group B was an increase in fitness measured (inferred from the relationship between blood lactate and velocity during standardised exercise tests). The increase in horse height in Groups B and C was greater than in Group A. Increases in girth and weight were smaller in Group A than in Groups B and C. Groups A and B had 62.5% and 12.5% of horses with locomotor diseases respectively. The authors concluded that a training program that includes swimming training could reduce locomotor diseases in young horses.

5.4. IMPORTANCE OF RECOVERY DAYS TO PROMOTE WELFARE OF HORSES IN TRAINING

During the 2 days after a race or intense sprint training, horses should be either completely rested, or only lightly exercised. It is during this period after fast exercise that the training responses actually occur. Cells are actively repairing damaged structures, and anabolism, or protein building, occurs. The horse also restores its muscle cell glycogen content over a 24 hour period.

Dysfunction of the lower respiratory tract has also been reported after intense exercise and transport (Raidal et al., 1997a, b). The implications of these observations for performance or the incidence or severity of disease in horses during strenuous training are unknown. However, avoidance of strenuous exercise during the 2–3 day recovery period after intense exercise is advisable.

During recovery days the appetite, gait and attitude of horses should be closely observed. Minor injuries should be attended to, and ice packs used on any areas that are inflamed (heat and swelling indicate inflammation). The flexible ice packs are excellent for strapping onto shins, fetlocks and flexor tendons for 20–30 minutes or so.

There are no special strategies for recovery from prolonged exercise except provision of water, electrolytes, and a high-energy diet. Prolonged washing of horses with very cold water will assist the cooling down process immediately after racing or competition. Low energy diets such as hay may contribute to delayed glycogen resynthesis in the 2–3 day period after exercise (Snow et al., 1987). This delay is of little consequence unless the horse is competing on successive days. There is no decline in muscle glycogen content with a training protocol typical of that used in many British thoroughbred training yards, and glucose supplements are unnecessary (Snow et al., 1991).

5.5. OVERTRAINING

Many horses have periods of short-term fatigue during their training. This fatigue is usually accompanied by reduced appetite. However, the horse usually recovers in 2–4 days. This short-term fatigue after racing or intensive training is more common in horses that have been inadequately prepared for the intense exercise.

Overtraining refers to a syndrome similar to chronic fatigue. It is associated with reduced performance that is not corrected by several weeks of rest. This more severe form of fatigue results in reduced body weight and loss of interest in exercise. The syndrome is caused by repeated use of high intensity training without adequate rest periods. If the training is not stopped there is a serious threat to the animal's welfare. It has been shown that the syndrome is associated with a reduction in the hormonal response to exercise. Golland et al. (1999) found that the cortisol response to a standardised treadmill exercise test was decreased in overtrained horses. This finding was not associated with evidence of adrenal exhaustion.

Regular monitoring of body weight is the best way to detect early signs of overtraining. Trainers may also detect signs of mood change in some horses in the early stages of this syndrome. Unfortunately there are no blood tests that can be performed in resting horses that predict the onset of overtraining syndrome.

In conclusion, the welfare of racehorses will be promoted by training that neither undertrains nor overtrains the horse. Properly prepared horses will be fit to compete in their first race, and should recover well after their first race. Excessive use of racing or competition as a training method is likely to result in fatigue and lameness, and high wastage rates. Use of heart rate measurements during exercise can help refine training and promote horse welfare.

6. References

Bailey, C.J. (1998) *Wastage in the Australian Thoroughbred Racing Industry*. Rural Industries Research and Development Corporation, Canberra, Australia.

Boston, R.C. and Nunamaker, D.M. (2000) Gait an speed as exercise components of risk factors associated with onset of fatigue injury of the third metacarpal bone in 2-year-old Thoroughbred racehorses. *American J. Veterinary Research* **61**, 602–608.

Bourke, J.M. (1995) Wastage in Thoroughbreds. *Proceedings of Annual Seminar of the Equine Branch of New Zealand Veterinary Association*, pp. 107–119.

Brunker, P., Bennell, K. and Matheson, G. (1999) *Stress Fractures*. Blackwell Science Asia, Carlton, Victoria, Australia.

Buckingham, S.H.W. and Jeffcott, L.B. (1990) Shin soreness: a survey of Thoroughbred trainers and racetrack veterinarians. *Australian Equine Veterinarian* **8**, 148–153.

Carrier, T.K., Estberg, L., Stover, S.M., Gardner, I.A., Johnson, B.J., Read, D.H. and Ardans, A.A. (1998) Association between long periods without high-speed workouts and risk of complete humeral or pelvic fracture in thoroughbred racehorses – 54 cases (1991–1994). *J. American Veterinary Medical Association* **212**, 1582–1587.

Cohen, N.D., Berry, S.M., Peloso, J.G., Mundy, G.D. and Howard, I.C. (2000) Thoroughbred racehorses that sustain injury accumulate accumulate less high speed exercise compared to horses without injury in Kentucky. *Proceedings 46th Annual Convention of the American Association Equine Practitioners* **46**, 51–53.

Davies, H.M.S. (1996) The effects of different exercise conditions on metacarpal bone strains in Thoroughbred racehorses. *Pferdeheilkunde* **12**, 666–670.

Davies, H.M., Gale, S.M. and Baker, I.D.C. (1999) Radiographic measures of bone shape in young Thoroughbreds during training for racing. *Equine Veterinary J. Suppl.* **30**, 262–265.

D'sterdieck, K.F., Schott, H.C., Eberhart, S.W., Woody, K.A. and Coenen, M. (1999) Electrolyte and glycerol supplementation improve water intake by horses performing a simulated 60 km endurance ride. *Equine Veterinary J. Supplement* **30**, 418–424.

Estberg, L., Gardner, I.A., Stover, S.M., and Johnson, B.J. (1998) A case-crossover study of intensive racing and training schedules and risk of catastrophic musculoskeletal injury and lay-up in California thoroughbred racehorses. *Preventive Veterinary Medicine* **33**, 159–170

Eaton, M.D., Evans, D.L., Hodgson, D.R., and Rose, R.J. (1995) Maximal accumulated oxygen deficit in thoroughbred horses. *J. Applied Physiology* **78**, 1564–1568.

Evans, D.L. and Walsh, J.S. (1997) Effect of increasing the banking of a racetrack on the occurrence of injury and lameness in Standardbred horses. *Australian Veterinary J.* **75**, 751–752.

Firth, E.C., Rogers, C.W. and Goodship, A.E. (2000) Bone mineral density changes in growing and training Thoroughbreds. *Proceedings of the 46th Annual Convention of the American Association of Equine Practitioners* **46**, 295–299.

Fredricson, I., Dalin, G., Drevemo, S., Hjerten, G. and Alm, L.O. (1975a) A biotechnical approach to the geometric design of racetracks. *Equine Veterinary J.* **7**, 91–96.

Fredricson, I., Dalin, G., Drevemo, S., Hjerten, G. and Alm, L.O. (1975b) Ergonomic aspects of poor racetrack design. *Equine Veterinary J.* **7**, 63–65.

Hernandez, J., Hawkins, D.L. and Scollay, M.C. (2001) Race-start characteristics and risk of catastrophic musculoskeletal injury in Thoroughbred racehorses. *J. American Veterinary Medical Association* **218**, 83–86.

Geor, R.J., MsCutcheon, L.J. and Lindinger, M.I. (1996) Adaptations to daily exercise in hot and humid ambient conditions in trained Thoroughbred horses. *Equine Veterinary J. Suppl.* **22**, 63–68.

Geor, R.J. and McCutcheon, L.J. (1998) Hydration effects on physiological strain of horses during exercise-heat stress. *J. Applied Physiology* **84**, 2042–2051.

Golland, L.C., Evans, D.L., Stone, G.M., Tyler-McGowan, C.M., Hodgson, D.R. and Rose, R.J. (1999) Plasma cortisol and B-endorphin concentrations in trained and overtrained Standardbred racehorses. *Pflugers Archiv – European J. Physiology* **439**, 11–17.

Gottliebvedi, M., Persson, S., Erickson, H. and Korbutiak, E. (1995) Cardiovascular, respiratory and metabolic effects of interval training at vla4. *J. Veterinary Medicine – Series A* **42**, 165–175.

Gysin, J., Isler, R. and Straub, R. (1987) Evaluation of performance capacity and definition of training intensity using heart rate and blood lactate measurements. *Pferdeheilkunde* **3**, 193.

Jansson, A., and Dahlborn, K. (1999) Effects of feeding frequency and voluntary salt intake on fluid and electrolyte regulation in athletic horses. *J. Applied Physiology* **86**, 1610–1616.

Jeffcott, L.B., Rossdale, P.D., Freestone, J., Frank, C.J. and Towers-Clark, P.F. (1982). An assessment of wastage in Thoroughbred racing from conception to 4 years of age. *Equine Veterinary J.* **14**, 185–198.

Jeffcott, L.B., Buckingham, S.H., McCarthy, R.N., Cleeland, J.C., Scotti, E. and McCartney, R.N. (1988) Non-invasive measurement of bone: a review of clinical and research applications in the horse. *Equine Veterinary J. Suppl.* **6**, 71–79.

Jeffcott, L.B. and Kohn, C.W. (1999) Contributions of exercise physiology research to the success of the 1996 Equestrian Olympic Games: a review. *Equine Exercise Physiology 5, Equine Veterinary J. Suppl.* **30**, 347–355.

J.R.A. (1991) Preventing accidents to racehorses: studies and measures taken by the Japan Racing Association, *Report of the Committee on the Prevention of Accidents to Racehorses,* Japan Racing Association.

Kobluk, C.N., Geor, R.J., King, V.L. and Robinson, R.A. (1996) A case control study of racing Thoroughbreds conditioned on a high-speed treadmill. *J. Equine Veterinary Science* **16**, 511–513.

Lindner, A. and Dingerkus, A. (1993). Incidence of training failure among Thoroughbred horses at Cologne, Germany. *Preventative Veterinary Medicine* **16**, 85–94.

Lovell, D.K. and Rose, R.J. (1991) Changes in skeletal muscle composition in response to interval and high intensity training. *Equine Exercise Physiology* **3**, ICEEP Publications, Davis, Canada, pp. 215–222.

Marlin, D.J., Scott, C.M., Schroter, R.C., Mills, P.C., Harris, R.C., Harris, P.A., Orme, C.E., Roberts, C.A., Marr, C.M., Dyson, S.J. and Barrelet, F. (1996) Physiological responses in nonheat acclimated horses performing treadmill exercise in cool (20 °C/40%RH), hot dry (30 °C/40%RH) and hot humid (30 °C/80%RH). *Equine Veterinary Journal. Supplement* **22**, 70–84.

Mason, T.A. and Bourke, J.M. (1973). Closure of the distal radial epiphysis and its relationship to unsoundness in two year old Thoroughbreds. *Australian Veterinary J.* **49**, 221–228.

McKane and Slocombe (1999) Sequential changes in bronchoalveolar cytology after autologous blood inoculation, Equine Exercise Physiology 5. *Equine Veterinary J. Supplement* **30**, 126–130.

Meyer, T.S., Fedde, M.R., Gaughan, E.M., Langsetmo, I. and Erickson, H.H. (1998) Quantification of exercise-induced pulmonary haemorrhage with bronchoalveolar lavage. *Equine Veterinary J.* **30**, 284–288.

Misumi, K., Sakamoto, H. and Shimizu, R. (1994) The validity of swimming training for 2-year-old thoroughbreds. *J. Veterinary Medical Science* **56**, 217–222.

Moyer, W., Spencer, P.A. and Kallish, M. (1991) Relative incidence of dorsal metacarpal disease in young Thoroughbred racehorses training on two different surfaces. *Equine Veterinary J.* **23**, 166–168.

Moyer, W. and Fisher, J.R.S. (1992) Bucked Shins: Effects of differing track surfaces and proposed training regimes. *Proceedings of the 37th Annual Convention of the American Association of Equine Practitioners,* 541–547.

Munoz, A., Riber, C., Santisteban, R., Rubio, M.D., Aguera, E.I. and Castejon, F.M. (1999) Cardiovascular and metabolic adaptations in horses competing in cross-country events. *J. Veterinary Medical Science* **61**, 13–20.

Murakami, M., Imahara, T., Inui, T., Amada, A., Senta, T., Takagi, S., Kubo, K., Sugimoto, O., Watanabe, H., Ikeda, S. and Kameya, T. (1976) Swimming exercises in horses. *Experimental Reports Equine Health Laboratory* **13**, 27–48.

Nielsen, B.D., Potter, G.D., Morris, E.L., Odom, T.W., Senor, M.A., Reynolds, J.A., Smith, W.B. and Martin, M.T. (1997) Changes in the third metacarpal bone and frequency of bone injuries in young quarter horses during race-training- observations and theoretical considerations. *J. Equine Veterinary Science* **17**, 541–549.

Nielsen, B.D., Potter, G.D., Greene, L.W., Morris, E.L., Murraygerzik, M., Smith, W.B. and Martin, M.T. (1998) Response of young horses in training to varying concentrations of dietary calcium and phosphorus. *J. Equine Veterinary Science* **18**, 397–404.

Norwood, G.L. (1978) The bucked-shin complex in Thoroughbreds. *Proceedings of the 24th Annual Convention American Association Equine Practitioners,* pp. 319–336.

Nunamaker, D. (1996) Stress fractures in Thoroughbred racehorses. *Surgery Forum,* 1–4.

Physick-Sheard, P.W. (1986) Career profile of the Canadian Standardbred I. Influence of age, gait and sex upon chances of racing. *Canadian J. Veterinary Research* **50**, 449–456.

Pickersgill, C.H. and Reid, S.W.J. (2002a) Musculoskeletal injuries and associated epidemiological risk factors among Thoroughbred racehorses in the UK, in press.

Pickersgill, C.H., Marr, C.M. and Reid, S.W.J. (2002b) The epidemiology of superficial digital flexor tendinitis among National Hunt racehorses in the UK, in press.

Pool, R.R. (1991) Pathology of the metacarpus. *Proceedings of the 13th Bain-Fallon Memorial Lectures.* Australian Association Equine Practitioners, Sydney, pp. 105–117.

Pool, R.R. and Meagher, D.M. (1990) Pathologic findings and pathogenesis of racetrack injuries. *Veterinary Clinics North America: Equine Practice* **6**, 1–30.

Raidal, S.L., Love, D.N. and Bailey, G.D. (1997a) Effect of a single bout of high intensity exercise on lower respiratory tract contamination in the horse. *Australian Veterinary J.* **75**, 293–295.

Raidal, S.L., Bailey, G.D. and Love, D.N. (1997b) Effect of transportation on lower respiratory tract contamination and peripheral blood neutrophil function. *Australian Veterinary J.* **75**, 433–438.

Riggs, C.M. and Evans, G.P. (1990) The microstructural basis of the mechanical properties of equine bone. *Equine Veterinary Education* **2**, 197–205.

Robinson, R.A. and Gordon, B. (1988). American Association of Equine Practitioners track breakdown studies – horse results. *Proceedings 7th International Conference of Racing Analysts and Veterinarians*, pp. 385–394.

Ross, W.A. and Kaneene, J.B. (1996) An operation-level prospective study of risk factors associated with the incidence density of lameness in Michigan (USA) equine operations. *Preventative Veterinary Medicine* 28, 209–224.

Rossdale, P.D., Hopes, R., Wingfield Digby, N.J. and Offord, K. (1985) Epidemiological study of wastage among racehorses 1982 and 1983. *Veterinary Record* **116**, 66–69.

Schamhardt, H.C., Merkens, H.W. and van Osch, G.J.V.M. (1991) Ground reaction force analysis of horses ridden at the walk and trot. *Proceedings of the 3rd Equine Exercise Physiology Conference* **3**, 120–127.

Schott, H.C., McGlade, K.S., Molander, H.A., Leroux, A.J. and Hines, M.T. (1997) Body weight, fluid, electrolyte, and hormonal changes in horses competing in 50- and 100-mile endurance rides. *American J. Veterinary Research* **58**, 303–309.

Schroter, R.C., Marlin, D.J. and Denny, E. (1998) Exercise-induced pulmonary haemorrhage (EIPH) in horses results from locomotory impact induced trauma – a novel, unifying concept. *Equine Veterinary J.* **30**, 186–192.

Sinha, A.K., Ray, S.P. and Rose, R.J. (1991) Effect of training intensity and detraining on adaptations in different skeletal muscles. *Equine Exercise Physiology* **3**, ICEEP Publications, Stockholm, 223–230.

Sosa, Leon, L.A., Hodgson, D.R., Carlson, G.P. and Rose, R.J. (1998) Effects of concentrated electrolytes administered via a paste on fluid, electrolyte and acid base balance in horses. *American J. Veterinary Research* **59**, 898–902.

Snow, D.H., Harris, R.C, Harman, J.C. and Marlin, D.J. (1987) Glycogen repletion following different diets. *Equine Exercise Physiology* **2**, ICEEP Publications, California, pp. 701–710.

Snow, D.H. and Harris, R.C. (1991) Effects of daily exercise on muscle glycogen in the thoroughbred racehorse. *Equine Exercise Physiology* **3**, ICEEP Publications, Stockholm, pp. 299–304.

Stover, S.M., Pool, R.R., Morgan, J.P., Martin, R.B. and Sprayberry, K. (1988) A review of bucked shins and metacarpal stress fractures in the Thoroughbred racehorse. *Proceedings of the 34th Annual Convention of the American Association of Equine Practioners* **34**, 129–136.

van Weeren, P.R., Brama, P.A.J. and Barneveld, A. (2000) *Proceedings of the 46th Annual Convention American Association Equine Practitioners* **46**, 29–35.

Watkins, K.L. (1985) A survey of pre-race veterinary withdrawals and post-race veterinary findings during one racing season in Hong Kong. *Proceedings 6th International Conference of Racing Analysts and Veterinarians*, pp. 343–346.

West, J.B. and Mathieucostello, O. (1994) Stress failure of the pulmonary capillaries as a mechanism for exercise induced pulmonary haemorrhage in the horse. *Equine Veterinary J.* **26**, 441–447.

Williamson, L.H., White, S., Maykuth, P., Andrews, F., Sommerdahl, C. and Green, E. (1995) Comparison between two post exercise cooling methods. *Equine Veterinary J. Supplement* **18**, 337–340.

Chapter 9

SPECIFIC WELFARE PROBLEMS ASSOCIATED WITH WORKING HORSES

R.T. WILSON
Bartridge Partners, Bartridge House, Umberleigh, North Devon EX37 9AS, UK

Abstract. A brief review of the history and uses of workhorses is provided as well as numbers and distributions of working equines at the beginning of the twenty-first century. The problems of using the correct harness and correct alignment of horse and equipment through the harness are considered and care of the feet and especially the necessity of correct shoeing for both welfare and maximum work are discussed. The nutritional needs of working equines in terms of energy, protein, minerals and vitamins and the importance of good health and prevention and management of disease, are examined. Relationships between the animal and his owner or handler and some legal aspects of improving welfare are covered and it is concluded that education of these owners/handlers is essential for improving the welfare of working equids in developing countries.

1. Introduction

Horses – and their congeneric the ass and the crosses between them that are the mule and the hinny – have served man as sources of energy for millennia. Their support roles have been in the military, in civil society in general, in industry and in agriculture. It is perhaps in their most readily recognisable work functions as traction and draught animals in industry and agriculture that they have made their greatest contribution to human welfare and advancement. High profile roles in industry in the now developed world ranged from that of the tiny pit pony often less than 9 hands (91 cm) at the withers and weighing as little as 20 stone (125 kg), to the brewery dray horse of more than 18 hands (185 cm) and weighing at the very least a ton (1000 kg) and often considerably more. In agriculture, plough teams were the image of rural production in Europe and especially the UK in the nineteenth century.

In western industry and agriculture, however, the use of the horse as the main source of energy following its demise from being the Great Horse of noble knights, to a mere tool of economic production was short lived. It really spanned no more than 250 years and may have been as short as 150 years from approximately 1760 to the 1920's after the 'golden age' of 1901–1914 and the 'Great War boom' of 1914–1920 (Chivers 1976). Superseding oxen on farms at the end of the seventeenth century, horses began to be replaced during the nineteenth century by the steam engine and then from the beginning of the twentieth by the more mobile and flexible internal combustion engine and the electric motor.

By 1950 horses as sources of energy had all but disappeared in the more advanced

N. Waran (ed.), The Welfare of Horses, 203–218.
© 2002 *Kluwer Academic Publishers. Printed in the Netherlands.*

western countries with larger scale industry and agriculture. They were, however, a considerable force in some countries with widespread peasant agriculture and in the planned economies of the socialist countries. In parts of the developing world and especially in countries such as Ethiopia and Mexico they are still important as sources of agricultural power. In other parts of the developing world and in some 'emerging' countries, horses often take second or third place to cattle and people, as sources of agricultural energy.

In both Ethiopia (FAO 2000) and in Mexico (Leon *et al.*, 2000) as well as in many others the numbers of horses at the turn of the twenty-first century, were greater than they had been for several decades. A remarkable phenomenon, however, is the resurgence of horse power in the developed world as sections of the community have recognised that fossil energy is a finite and limited resource and that the energy that horses can provide is not only sustainable through natural breeding, but can also be more sympathetic to the environment (see Miller 2000).

The resurgence of the working horse in the western world or at least a nostalgia for it, is exemplified by the plethora of recent books devoted to it (Miller 2000; Telleen 2000; Zeuner 2000). The proliferation of international conferences, seminars and training courses is also testimony to this interest. Further evidence of modern interest in working equines and especially in their welfare is in the number of charitable foundations working to protect horses and donkeys including The International League for the Protection of Horses, The Donkey Sanctuary and the Society for the Protection of Animals in North Africa (SPANA).

It is likely that welfare problems are most acute in the large areas of the developing world where small farmers with limited access to resources struggle to make their own living. Such people are disinclined or unable to devote time and money to providing an adequate quality of life to their animals. This situation has given rise to several 'development' projects devoted to improving the welfare of developing world equines by the foundations mentioned in the preceding paragraph and by other organisations such as FARM-Africa (a UK-based international Non-Government Organisation, (NGO)) and the World Association for Transport Animal Welfare and Studies.

2. Numbers And Distribution Of Working Horses

According to the database of the Food and Agriculture Organization of the United Nations (http://www.fao.org 2000) there were some 58.8 million horses in the world in 2000. Most were in Central and South America (Mexico 6.2 million, Brazil 5.9 million, Argentina 3.6 million, Columbia 2.5 million, Peru 0.7 million, Chile 0.6 million and Uruguay, Venezuela and Bolivia 0.5 million each). Other countries with large horse populations include China (8.9 million), Mongolia (3.1 million), Ethiopia (2.7 million), the Russian Federation (1.7 million), India (1.0 million), Kazakhstan (0.9 million) and Kyrgyzstan (0.3 million). In the developed world the horses of the USA (5.3 million) outnumber those of all other countries combined, with Germany having 476,000, Canada 385,000, France 349,000, Italy 280,000, Spain

248,000 and the United Kingdom 173,000 of which only 8000 are used for work (Spedding 2000).

Almost 80 per cent of all equines are located in the developing world. Some 97 per cent of all the Earth's mules are found in this economic grouping as are 96 per cent of donkeys and 60 per cent of horses. Some of the former socialist countries of Eastern Europe and the former Soviet bloc (Romania 842,000; Poland 750,000; Ukraine 698,000; Belarus 221,000; Bulgaria 141,000; Yugoslavia 76,000; Hungary 65,000) continue to have large numbers of horses compared to developed countries of similar geographical area or human population numbers (the large number of horses in Germany reflects the former division into the capitalist former Federal Republic of the 'west' and the socialist Democratic Republic of the 'east').

Almost all horses can be considered to perform work. Even those that are kept and ridden for pleasure in the western world expend energy in motion and load bearing that conforms to the definition of work. Most 'work' horses (those used for purposes other than recreation and sport) are, however, clearly to be found in the developing world. It is perhaps here that nutrition, health and welfare problems are likely to be of the greatest importance.

3. Nutrition

3.1. OVERVIEW

Visual assessment of the body condition of a working horse provides the best indication of its nutritional needs and especially of its nutritional status (Jones *et al.*, 1989). An exact knowledge of weight is of less use because feed needs of mature and near-mature working horses are related more to the intensity and duration of the tasks that they are called upon to perform than on theoretical needs correlated to weight. In spite of this, however, it is usual to calculate dietary requirements on the basis of weight (Table 1). The ideal in any nutritional regime should be to attain and maintain the ideal body condition for the expected tasks.

The quantity and quality of the diet are subjectively assessed as adequate if the horse has a sleek (not fat or thin) appearance, the coat shines and work demands are performed with vigour and apparent eagerness. In contrast the diet – any health

Table 1. Daily nutrient requirements for a mature horse weighing 500 kg.

Work regime	Dietary component and amount			
	Energy (mcal)	Protein (g)	Calcium (g)	Phosphorus (g)
Maintenance	16.4	656	20	14
Light work	20.5	820	25	18
Moderate work	24.6	984	30	21
Heavy work	32.8	1312	40	29

problems apart – almost certainly needs adjustment if the animal is fat or thin, the coat is dull and the attitude to work is reluctant.

The equine digestive system is adapted to almost continuous eating. Under free ranging conditions horses spend 12 or more hours a day in direct feeding activities. In doing this they seemingly satisfy a need to feel satiated while concurrently contributing to efficient digestion. A diet of high quality pasture would fulfil not only the physical but also the physiological nutritional requirements of most horses. As work is demanded of them, however, the time they are allowed to spend feeding is reduced. In addition feeding activities may be compressed into a shorter period. Supplementary feeding with high quality roughage or with concentrate feeds is therefore needed during breaks in work if the horse is not to suffer welfare problems which may in turn lead to reduced economic performance.

In concluding this overview it must be understood that poor nutrition – including inadequate access to water – leads to a whole range of other problems (Jeremy 1998). These include not only the direct effects but also reduced resistance to disease and an inability to perform work without undue stress (see Chapter 3; Nutrition and Welfare).

3.2. ENERGY REQUIREMENTS

An amount of 1.5–2.0 per cent of live body weight of good quality grass dry matter – as green fodder, hay or haylage – is generally sufficient to meet the daily maintenance energy needs of a horse (Orr *et al.*, 1989). The amount will vary slightly depending on the animal's particular basal metabolic rate and the amount of restless energy expended in activities such as walking, stamping, pawing and other displacement behaviour. Some of this behaviour may be caused by anxiety and may indicate a welfare problem or potential problem.

The need for energy increases with the amount of work. A horse that is underfed will not work well and certainly not nearly to its full potential. Dietary energy content is important because energy expenditure per unit time of aerobic work is probably exponentially related to the work demand (see Pearson 1998).

Coarse or fibrous feed – 'forage' – is essential to correct functioning of the digestive tract. It is possible that light work improves the digestibility of forage as well as increasing the voluntary intake. Diets in which the forage content is low increase the probability of digestive upsets and some associated behavioural problems such as wood chewing, 'cribbing' and 'weaving' (see Chapter 5). Some of the energy demands of work can be overcome by more frequent feeding. Forage should be fed in sufficient quantity to maintain a high level of intestinal fill which in turn provides reserves of water and electrolytes to replace those lost by sweating. Each kilogram of ingested forage dry matter results in a flow of 8–14 litres of water containing 135 mmol/litre of sodium and 10 mmol/litre of potassium into the caecum (Cunha 1991). Horses fed high levels of forage thus have high reserves of water and electrolytes. Forage is also a good source of the volatile fatty acids that are used as a substrate for the provision of energy ('work').

As the workload increases in time or in effort – or as a combination of the two

– more feed with a higher energy density than forage must be fed if performance is to be maintained. Cereal grains provide more energy substrate per unit of volume than does forage. As the weight of a cereal grain is related to fibre content bulky high fibre feeds weigh less per unit volume than low fibre feeds. In this context wheat or barley thus have higher energy densities than oats and it is important to feed grains by weight rather than by bulk. Grain should be introduced gradually to any diet to prevent digestive upsets (Cunha 1991).

High intensity anaerobic work such as heavy draught or pulling requires an increase in grain feeding to prevent a reduction in body condition. The energy required for strenuous work may be as much as 70 times that required for normal walking with no load (Hintz & Schryver 1987). Intense work may require the feeding of 1.4–2.3 kg of grain for each hour of activity. During prolonged periods of heavy work such as at spring ploughing or autumn harvesting it is possible that a horse may not physically be able to eat such amounts of feed. Even more energy dense feeds such as fats may then have to be substituted for part of the forage and grain components of the diet. Horses can consume diets that contain as much as 30 per cent fat and can metabolise as much as 85 per cent of the energy in some oils and fats. Most fats and oils are also highly palatable and readily appreciated by horses. As for grain, fats should be introduced gradually to the diet.

Muscle glycogen concentrations can be altered by the amount of work and manipulation of the diet. This can result in increased amounts of stored glycogen in skeletal muscles as animals adapt to the regime (Perez *et al.*, 1996). As glycogen can be used rapidly during heavy work a stored source delays the onset of fatigue. In all cases, the amount and quality of feed should be adjusted to maintain the desired body condition.

3.3. PROTEIN REQUIREMENTS

Protein requirements beyond those for maintenance are low if the energy needs of a working horse are met. Demand does increase to some extent with work but this is likely to be met if there is 8–10 per cent of crude protein in the overall diet and feed dry matter intake is at 2–3 per cent of body weight.

Feeding excessive amounts of protein can be detrimental to the horse (Glade 1984) as well as resulting in higher feed costs. Among the problems associated with feeding high levels of protein are: increased sweating, faster heart rates, faster respiration, increased urine output and higher levels of urinary nitrogen and higher drinking water needs.

3.4. MINERALS

Mineral needs for most horses are met by most feeding regimes. Where there is any doubt, however, the whole diet as fed or eaten should be analysed.

Mature horses have large reserves of calcium in their bones so that an increase in need due to work is readily accommodated. Calcium to phosphorus ratios are not usually a problem in mature horses, but may be, in young animals. Diets high in phosphorus – those having large amounts of maize or oats are in this category – should be avoided in growing stock as high phosphorus intakes can interfere with absorption of calcium.

Sodium, potassium and chloride are essential in maintaining fluid balance and for nerve and muscle function. These three electrolytes and magnesium, calcium and phosphorus are lost in increasing quantities as animals sweat under load. Sodium demand is particularly high in working horses as a result of sweating and performance will fall off rapidly if lost electrolytes are not replaced.

Selenium requirements increase proportionately with exercise. Iron – which is important for the synthesis of haemoglobin – is a component of sweat and is therefore lost when the horse sweats profusely. Iron is also lost through absorption in the gastrointestinal tract. In addition, iron needs may be higher in the early stages of work because of the increased rate of synthesis of red blood cells.

Free access to salt or a salt lick will usually solve any problems if diets are inadequate in sodium. Some additional supplementation with phosphorus may be required for diets that contain little cereal. Selenium should not be fed to excess as it may be toxic at high levels (Orr et al., 1985; Shallow et al., 1985). Iron is usually present in adequate quantities in most diets.

3.5. VITAMINS

Vitamin A is usually obtained in adequate quantities from green forage. The B vitamins derive directly from the diet or by microbial synthesis in the intestine. B-complex vitamins are important in red blood cell synthesis and in energy metabolism and are key components of working horse diets. The role of Vitamin C is not clear in horses but adequate amounts are synthesised from glucose in the liver. Vitamin D is synthesised from dry fodder and is produced by the effect of ultraviolet radiation on the skin. Vitamin E is a biologic antioxidant that protects membranes from damage.

More research is required on the need for and effects of vitamins in working horses. In the meantime where there are doubts as to an adequate supply proprietary supplements should be mixed in the feed.

4. Health

4.1. OVERVIEW

It has been said (Freeland 1991) that the commonest [health] conditions in working equines (with special reference to the tropics) are:

lameness, lameness, worms and lameness

that is to say

  – lame but still able to work

- lame but unable to work
- full of worms
- lame because of worms

followed by

back sores, sarcoids and other skin conditions.

Horses can become lame for a variety of reasons, some of the problems experienced by the sports horses described in Chapter 2, will also affect the working equine in under-developed countries. However in the tropics, lameness is frequently a result of poor quality hooves. Good hoof quality starts with good nutrition in combination with regular and appropriate hoof care. The deleterious effects of helminth parasites have been documented by various authors (see Clayton 1986; Klei 1986), and are also covered in Chapters 2 and 4. Despite this there is very little empirical information on parasite infection from large-scale studies of working equines in the tropics and sub-tropics. General conclusions are drawn from the smaller studies that have been carried out (*e.g.* Pandey & Eysker 1991), and what is already known about the prevalence of different species of parasites in the temperate region (Sewell 1991). Indeed, Hammond and Sewell (1990) concluded that there is general similarity in the species of parasites present in a wide range of climatic zones. Parasitism can result in impaired growth and lack of or reduced resistance to infectious disease. Heavy parasitism in nutritionally stressed and over-worked animals is therefore a main cause of weakness and low productivity and is obviously of welfare concern. A further welfare concern is the clinical problems caused by poorly designed harnesses and equipment and the over-loading and poor treatment of horses. These include various skin conditions, sores and sarcoids (see Freeman 1991). Many of these problems can be prevented through the use of better methods for harnessing and loading and through education of the owners/handlers of the horses.

4.2. WELFARE, FINANCIAL AND ECONOMIC CONSIDERATIONS

As already indicated under the section on nutrition, one of the best indicators of a horse's well being is its body condition. For working horses, a high condition score indicates not only that the animal is being fed correctly (although note that in the developed world, high scores can also indicate that the horse is being fed too much) but also that its underlying health is in a satisfactory state. A low score suggests that the animal is not eating properly or is not efficiently converting the feed it eats to serve its bodily functions. The latter situation is largely symptomatic of a health problem even in the absence of overt signs of disease.

It is natural for most people to be prepared to spend money on visible disease or conditions (Pearson 1998). These include lameness, saddle sores, some external parasites and the skin diseases referred to at the beginning of this section. There is usually much more resistance to paying for treatment for invisible diseases such those caused by internal intestinal parasites and blood protozoa. In the latter category a particular problem in horses in some parts of the tropics is trypanoso-mosis. This debilitating disease can often pass undetected for long periods as it is not always easy to identify the parasites in the blood but it causes considerable welfare problems in addition to financial loss.

In general many horse owners tend to treat health problems at a late stage of development rather than in the early stages. Here as in many other aspects of life prevention is better than cure. Prevention is clearly good for welfare but a further aspect is the purely financial one. In many parts of the world where horses are important as working animals the cost of a single dose of some drugs may be equal to or exceed the value of the horse itself.

Because of the high and even excessive cost of drugs to small-scale farmers with limited resources, alternative methods to improve health and welfare should be adopted. For internal parasites, for example, these involve adapting management to the parasite life cycle. If animals are housed, faeces can be removed regularly to interrupt parasite life cycles. For grazing animals frequent rotation of pastures and of the domestic animal species mix serves the same purpose (Herd 1986). Biological control may become a realistic option in the future through the use of nematophagous fungi and other parasite predators (Bird & Herd 1995). Traditional measures to control worms (such as certain plants) are used in much of Africa (Bizimana 1994). Alternative worm control measures may not only be good for welfare but also for the environment, as they will help to reduce the incidence of drug resistance to many pharmaceutical products.

Grooming
Horses living naturally and on grazing alone seldom need grooming. As long as the horse lives in a near natural state and consumes an essentially laxative diet the remains of the feed ingested and the excretions through the system are evacuated mainly by the action of the kidneys and the intestines. Concentrate feed and work that causes sweating alter the natural balance and waste products are secreted to a very considerable extent through the skin. These products clog the skin pores and prevent its natural breathing action. Cleansing has to be carried out to clear away the waste products and allow the skin to function normally. Regular and thorough grooming is thus essential to the health and welfare of working equines and should not be regarded as something that is merely cosmetic and carried out to improve an animal's appearance.

4.3. CONCLUSIONS ON HEALTH

Most conditions related to poor health or disease can be anticipated and can be prevented or controlled relatively easily. Improved husbandry and better care of the animal, its feeding, its work and above all its rest are essential to good welfare. It is thus unfortunate that in the areas where most horses are found their owners cannot or will not afford the extra time, money and attention that is needed to overcome these problems. Often the owner may be unaware that there is a potential or actual problem and therefore a possible solution. It is also in the countries where there are most horses that there is least public awareness of welfare issues and least legislation to encourage improvement of welfare.

5. Harness, Harnessing And Draught Efficiency

5.1. HARNESS TYPES

Harness is the gear carried by a horse that connects it to an implement or other type of equipment and through which the force generated by the animal is turned into productive work. Other than the minor pieces of equipment that constitute the complete set of harness, the two major ways in which horses are harnessed to permit use of their force are by a collar or a breastband. Horses can transfer more of their force through collars than breastbands.

The collar has been traditionally used in western societies for most operations including pulling wheeled vehicles, driving stationary equipment and for agricultural operations. Collars are expensive to make and need to be tailored more closely to the animal than a breastband if they are not to cause welfare problems and reduce performance efficiency. It is perhaps partly because of the expense and the need for fitting that the breastband is favoured over the collar method in Africa, Asia and Latin America.

5.2. PROBLEMS WITH HARNESS AND EQUIPMENT

There are often major deficiencies in the ways animals are harnessed, causing suffering in addition to the more obvious effect of preventing the most effective and economical use of the animal's strength. The topic is most pronounced and important in developing countries due to the large number of animals involved, the extent of the suffering and the overall economic importance of the work performed.

The general tendency in the developing world, even with the cheaper breast harnesses, is to go for least cost. The bands themselves are often too thin and lack padding. Bands and fastenings made of unsuitable materials are common (Pearson 1998). Repairs to damaged equipment are usually of a perfunctory nature and cause further abrasion of the animal's skin. In most areas the improved harness designs that cause fewer health and welfare problems need to cost as little as the badly designed and maintained equipment, if they are to be acceptable. Achieving this goal may not be too difficult.

A problem with 2-wheeled carts with many traditional types of harness is that much of the vertical force of the load is on the animal's neck rather than on its back. The use of harness designed for oxen or other animal species can also result in a similar situation. A suitable saddle with ropes or chains to hold the shafts or central pole easily overcomes this problem. In a converse situation loading the cart behind its wheels results in pressure on the horse's belly (Figure 1). Any work saddle should be suitably padded and fit well over the animal's back. Similar precautions should be taken with pack saddles.

Figure 1. Cart harness in Viet Nam (note 'breastband', cart loaded behind centre of gravity and lack of breeching straps for braking).

In addition to the neck and back sores that are almost universally seen there are often similar problems around the hocks on the back legs. These arise from inefficient braking systems on many carts. Brakes are not always necessary but if they are absent it is essential that proper breeching straps are part of the harness for shafted equipment. A front swingle or pole to the horse's collar should be used in pole carts.

5.3. THE APPLICATION OF POWER

Efficient power application has major welfare implications as it reduces the onset of fatigue. Draught efficiency is increased if the force exerted is close to the horizontal plane. The most usual methods of reducing the angle of pull are to lengthen the chain or other type of trace between the animal and the equipment and to lower the hitch point on the animal (*i.e.* on its collar or breastband) (Figure 2). Lower hitch points allow the centre of gravity to be moved downwards and forwards which contribute to more effective use of the horse's own weight (Hobbs 2000) in addition to any strength applied. A further method of reducing the angle of pull, is to use a hip strap through which the traces pass before descending to the implement being pulled (Inns 1991).

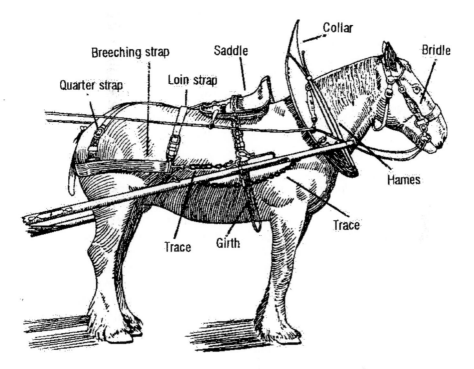

Figure 2. Main items of single cart harness (Source: Miller and Robertson 1959).

5.4. CONCLUSIONS ON HARNESS AND EQUIPMENT

With respect to efficient harness promoting good animal welfare it should (Barwell & Ayre 1982):

- Suit the principal physical characteristics of the animal with which it will be used;
- Be designed to ensure that the muscles are effectively used and that an animal's full weight is put into the work;
- Not cause discomfort or injury and so have large padded areas, be securely fastened to the animal and be capable of being adjusted to changes in the animal's size or body condition in order to prevention friction rubbing; and
- Be designed so that the draught force is as horizontal as possible.

In order that the implements fit the characteristics of the harness they should:

- Have suitable attachment points;
- Be sufficiently flexible to allow them to be manoeuvred easily where the situation requires it (*i.e.* trailed implements to be turned at field ends and wheeled equipment to be reversed – which also requires the correct harness);

- Be designed to prevent them running into the animal and allow animals to apply a braking force;
- Be designed so that downward loads on the animal are supported efficiently, that there is no upward load under the belly and so that 2-wheel carts cannot tip backwards; and
- Have some vertical flexibility to allow for uneven ground between the horse and the tool.

6. Hooves, Hoof Care And Shoeing

The structure and functioning of the hoof underline its role as the protective covering of the foot (Dollar 1898). In wild and many free living horses the protection is near perfect and it is rare to see bad feet in such animals because the bearing surface is renewed from above as it is worn away. Almost invariably and almost immediately that a horse is called upon to draw loads or do other types of work, however, the wear exceeds the rate of replacement. Some kind of protection of the bearing surface then becomes indispensable. The most common protector is the iron shoe although shoes of other material are occasionally used.

Much of the lameness referred to under the section on health is due to injury or insults to the foot and the coronary band immediately above it. Not a little of this is caused by poor hoof trimming and shoeing practices. The essential of all good shoeing is the preservation of the foot in a sound and healthy state. The farrier should be considered as a scientist as well as an artist (Fitzwygram 1903). He is a scientist because to be able to fit shoes presupposes an intimate knowledge of the processes at work in hoof renewal and repair. Improving faulty hooves and partially or completely restoring the function of diseased ones are important aspects of this science. He is an artist because he is a craftsman engaged not only in the making of shoes but also in fashioning them to the foot and to the ground surface and to ensuring secure and proper fitting so that the limb is always in the correct position.

The purposes of shoeing are thus manifold. Shoes prevent excessive hoof wear. They protect the sensitive structures, which the hoof contains. Slipping on roads or in muddy fields is reduced by fitting the correct shoes for the job. Faulty action such as 'dishing' (*i.e.* toes turned in) can similarly be corrected by a skilled farrier using specially adapted shoes. Diseased hooves can be protected at least in part by proper shoes. All these factors have welfare implications as they reduce suffering and pain whilst also allowing the horse to perform his expected functions satisfactorily. Even the less sensitive owners may be persuaded to provide better welfare for horses that are seen to be performing well and costing less to keep than animals with lower output.

Major problems in foot care and shoeing remain in many developing countries. Although in some areas such as Ethiopia some type of 'shoe' (often a piece of old motor car tyre) is used, it is often badly fitted and the nails may pierce the sensitive part of the foot. The main causes of lack of concern for foot care are lack of perception by and education of the owners who are frequently in pain themselves and see no obvious solutions. Some efforts are being made especially by some of the NGOs (non government organisations) already mentioned and by the Brooke Hospital for Animals to help owners to understand welfare aspects through support for development and for local adaptive research.

7. The Animal And His Owner: Training And Work

7.1. OVERVIEW

Animals need to be trained before they can be used for work (Spedding 2000). Any standards applied must therefore refer to both aspects but the basic ethical proviso is that no harm should be done (Fraser 1992). There are, however, few standards and in any case much depends on human attitudes. The general conclusions of a symposium on animal training held in 1990 (UFAW 1990) were that:

- Most animals appear to enjoy being trained;
- Training depends on the animal's co-operation and is best based on the use of rewards;
- Punishment, other than by voice, should only be used where an animal could endanger itself or others;
- Good welfare for even the most exploited animals (such as guide dogs) depends largely on a good environment outside their working time;
- Training is best regarded as an art rather than a science; and
- Trainers themselves should be trained.

7.2.   THE TRAINER

The best trainers are those who show deliberate and economical movements, exhibit quiet patience and use the least amount of equipment (Fraser 1992). Perhaps less easily seen but of equal or greater importance is that their concern for the horse transcends the basic items of food, water and shelter. They are concerned with behaviour and the facilitation of a broad range of behaviour.

Successful 'breaking' – clearly an inappropriate but still accepted term in the context of horse training – is best considered as the transfer of the horse's dependence from its own kind to its handler and to people in general. The best trained horses are those that are brought to realise that their trainer is a companion who can be trusted (Cregier 1986). Such horses seek comfort by approaching their handlers when faced with a threat rather than taking flight. It is the horse's comparative reactivity, sensitivity and need for companionship that make it one of the domestic animals most capable of bonding with its handler. Unfortunately it also renders it the most susceptible to the greatest cruelties. Such cruelties are most often inflicted on horses in developing countries where, as already indicated in the section on hoof care and shoeing, owners themselves often suffer almost constant pain and lack exposure to welfare issues through formal or non-formal education via the media.

7.3. TRAINING

Many of the principles of training are based on conscious modification of innate behaviour. A horse is required to perform many acts which are unnatural to it. It must, for example, learn to give up its feet for inspection and shoeing even when it fears they may be trapped. It must allow a heavy weight on its back when all its natural instincts tell it that this is a predator come to do harm (methods of training are covered in detail in Chapter 7).

Many of these innate inclinations can best be overcome or modified as soon after the birth of a foal as possible. Early handling is based on the concept that the young animal is better equipped than the adult to absorb and retain social experiences. Newborn foals that are intensively handled immediately on birth can be induced to accept much of what later would be considered to be very frightening. This applies to all of visual, auditory and tactile stimuli that it will or may experience during its association with people.

Even at older ages, however, the approach to training should focus on preventing or mitigating adverse tension between the trainer and the animal being trained. The companion-seeking or affiliative behaviour of the horse enhances its ability to mimic and to learn.

To be most effective, training should be sequential. Foals cannot be expected to do more than be led, tied and stand. Progress for workhorses should be very gradual. Little work should be demanded until a horse is three years old. At this stage it may be introduced to harness and loaded very lightly, preferably in close association with a fully trained companion. At four years the horse is ready for more serious and longer periods of work. Larger workhorses are not fully mature until at least five years of age and possibly older and it is only at this stage that prolonged hard work should be demanded. Horses that have benefited from such welfare considerations and that continue to benefit from sympathetic treatment can be expected to have a working life in excess of 10 years.

The main problems in adopting the precepts of training are evident in the developing countries. Horses are usually put to work at too young an age and are required immediately to perform the heaviest of tasks. The results are often stunted growth, susceptibility to disease and impaired reproductive performance.

7.4. LEGISLATION

There is a considerable amount of legislation in developed and in some developing countries covering animal welfare. Much of the legislation is being continually updated but not all is applicable to work animals. Some international conventions could act contrary to improved welfare. This would be the case, for example, where some countries are prevented from banning under the guise of free trade the import of animals from others countries with lower or no welfare standards. It is then difficult for the countries with high welfare standards to bring pressure to bear on those with poorer records.

8. Conclusion

The principle aspects of this chapter on the welfare of working horses are that:

* A broad range of military, civil society, industrial and agricultural functions have been undertaken by horses over a long period of time but their major service in the last two sectors lasted only about 250 years in the western world;
* At the beginning of the twenty-first century working horses are few in number and have only a minor role in the developed world but are extremely important numerically and economically in Latin America and parts of Africa and Asia where welfare conditions may be less than optimal;
* Adequate nutrition and good health are essential to good welfare and efficient work output, but these aspects can be problematic in some areas where working horses are most numerous and most work is required of them;
* The correct type of harness for the job to be performed is essential to good welfare and to contributing to efficient work output;
* Care of the feet and hooves and correct shoeing is required if welfare considerations are to be respected and animals are not to suffer unduly from the work expected of them;
* Training in which both handlers and horses should be intimately involved in a relationship of mutual trust and respect is a preliminary to animals being employed for work; and
* Legislation covering the use of animals for work and their welfare is better (and better respected) in some countries than in others and is still evolving, but some aspects of legislation relating to global trade may be inimical to improved welfare.

9. References

Barwell, I. and Ayre, M. (1982) *The Harnessing of Draught Animals.* Intermediate Technology Publications, London, UK.

Bird, J. and Herd, R.P. (1995) *In vitro* assessment of two species of nematophagus fungi (*Arthrobotrys oligospora* and *Arthrobotrys flagrans*) to control the development of infective cythostome larvae from naturally infected horses. *Veterinary Parasitology* **56**, 181–187.

Bizimana, N. (1994) *Traditional Veterinary Practice in Africa (Schriftenreihe der GTZ No. 243).* Deutsche Geselleschaft fur Technishe Zusammenarbeiten, Eschborn, Germany.

Clayton, H.M. (1986) Ascarids: recent advances, the veterinary clinics of North America. *Equine Practice* **2**, 313–328.

Chivers, K. (1976) *The Shire Horse: A History of the Breed, the Society and the Men.* J.A. Allen, London, UK.

Cregier, S.E. (1986) Horsebreakers, tamers and trainers: An historical, psychological, and social review. In Fox, M.W. and Lickley, L.D. (eds.), *Advances in Animal Welfare Science, 1986/1987.* Humane Society of the United States, Washington DC, USA, pp. 89–101.

Cunha, T.J. (1991) Feeding the high-level performance horse. In Cunha, T.J. (ed.), *Horse Feeding and Nutrition,* 2nd ed. Academic Press, San Diego, USA, pp. 331–348.

Dollar, J.A.W. (1898) *A Handbook of Horse-Shoeing with Introductory Chapters on the Anatomy and Physiology of the Horse's Foot.* David Douglas, Edinburgh, UK.

FAO (2000) *Production Yearbook*, Vol. 54. Food and Agriculture Organization of the United Nations, Rome, Italy.

Fitzwygram, F. (1903) *Horses and Stables*, 5th ed. Longman, Green, and Co., London, UK.

Fraser, A.F. (1992) *The Behaviour of the Horse*. CAB International, Wallingford, UK.

Freeland, G.G. (1991) Donkeys, mules and horses in tropical agricultural development. In Fielding, D. and Pearson, R.A. (eds.), *Donkeys, Horses and Mules in Tropical Agricultural Development*. University of Edinburgh, Edinburgh, UK, pp. 322–325.

Glade, M.J. (1984) Feeding innovations for the performance horse. *J. of Equine Veterinary Science* 4, 165–166.

Hammond, J.A. and Sewell, M.N.H. (1990) Diseases by Helminth. In Sewell, M.N.H. and Focklesby, D.W.B. (eds.), *Handbook on Animal Diseases in the Tropics*, 4th Edition. Bailliere Tindall, London, UK.

Herd, R.P. (1986) Epidemiology and control of equine strongylosis at Newmarket. *Equine Veterinary J.* 18, 477–452.

Hintz, H.F. and Schryver, H.F. (1987) Energy and protein. In Robinson, N.E. (ed.), *Current Therapy in Equine Medicine*, 2nd ed. W.B. Saunders Co., Philadelphia, 387-393.

Hobbs, S.J. (2000) Draught testing of a work horse. *Draught Animal News* 33, 2–4.

Inns, F.M. (1991) The design and operation of animal/implement systems: guidelines for soil cultivation implements. In Fielding, D. and Pearson, R.A. (eds.), *Donkeys, Horses and Mules in Tropical Agricultural Development*. University of Edinburgh, Edinburgh, pp. 258–265.

Jeremy, J. (1998) The nutrition of working horses in developing countries. In *3er Coloquio Internacional sobre Equidos de Trabajo*. Universidad Nacional Autonoma de Mexico, Mexico City, pp. 281–285.

Jones, R.S., Lawrence, T.J.L., Veevers, A., Cleeve, N. and Hall, J. (1989) Accuracy of prediction of the live weight of horses from body measurements. *Veterinary Record* 125, 549–553.

Klei, T.R (1986) Other parasites: recent advances, the veterinary clinics of North America. *Equine Practice* 2, 329–336.

Leon, A.C., Saldaña, T.M. and Miranda, C.R. (2000) La traccion animal en México. *Draught Animal News* 33, 5–9.

Miller, L.R. (2000) *Work Horse Handbook*. Rural Heritage, Gainseboro, TN, USA.

Miller, W.C. and Robertson, E.D.S. (1959) *Practical Animal Husbandry*, 7th ed. Oliver and Boyd, Edinburgh, UK.

Orr, E.A., Baker, J.P. and Hintz, H.F. (1989) Nutrient requirement tables. In Orr, E.A. (ed.), *Nutrient Requirements of Horses*, 5th ed. National Academy Press, Washington DC, USA, pp. 39–48.

Pandey, V.S. and Eysker, M. (1991) Internal parasites of equines in Zimbabwe. In Fielding, D. and Pearson, R.A. (eds.), *Donkeys, Mules and Horses in Tropical Agricultural Development*. University of Edinburgh, Edinburgh, UK, pp. 167–174.

Pearson, R.A. (1998) The future of working equids – prospects and problems. In *3er Coloquio Internacional sobre Equidos de Trabajo*. Universidad Nacional Autonoma de Mexico, Mexico City pp. 1–20.

Perez R., Valenzuela, S., Merino, V., Cabezas, L., Garcia, M., Bou, R. and Ortiz, P. (1996) Energetic requirements and physiological adaptation of draught horses to ploughing work. *Animal Science* 63, 343–351.

Sewell, M.N.H. (1991) Uniformity and contrasts of helminth diseases of equids. In Fielding, D. and Pearson, R.A. (eds.), *Donkeys, Mules and Horses in Tropical Agricultural Development*. University of Edinburgh, Edinburgh, UK, pp. 141–151.

Shallow, J.S., Jackson, S.G. and Baker, J.P. (1985) The influence of dietary selenium levels on blood levels of selenium and glutathione peroxidase activity in the horse. *J. Animal Science* 61, 590–598.

Spedding, C. (2000) *Animal Welfare*. Earthscan Publications Ltd., London.

Telleen, M. (2000) *The Draft Horse Primer*. Rural Heritage, Gainseboro, TN, UK.

UFAW (1990) *Animal Training (Proceedings of a Symposium)*. Universities Federation for Animal Welfare, Cambridge, UK.

Zeuner, D. (ed.) (2000) *The Working Horse Manual*. Farming Press, Ipswich.

INDEX